David Julian McClements
Food Nanotechnology

Also of interest

David Julian McClements

Food
Nanotechnology

—

DE GRUYTER

Author
Prof. David Julian McClements
Department of Food Science
University of Massachusetts
102 Holdsworth Way
Amherst MA 01003
United States of America
mcclemen@umass.edu

ISBN 978-3-11-078842-6
e-ISBN (PDF) 978-3-11-078845-7
e-ISBN (EPUB) 978-3-11-078854-9

Library of Congress Control Number: 2022942703

Bibliographic information published by the Deutsche Nationalbibliothek
The Deutsche Nationalbibliothek lists this publication in the Deutsche Nationalbibliografie;
detailed bibliographic data are available on the internet at http://dnb.dnb.de.

www.degruyter.com

Dedicated to my Daughter Isobelle and Wife Jayne

Acknowledgments

I thank my daughter Isobelle, my wife Jayne, and all my family and friends in the United Kingdom for all their support and encouragement throughout my career. I also thank all my innovative colleagues at the University of Massachusetts for creating such a stimulating and pleasurable environment to work. I also thank all the undergraduate students, graduate students, Post-Docs, and visiting scientists who have worked in my laboratory for their valuable insights and discussions. Finally, I thank Helene Chavaroche and Stella Muller from De Gruyter for their help in bringing this book from an initial concept to its final form.

https://doi.org/10.1515/9783110788457-202

Contents

About the author

David Julian McClements was born in the north of England but has lived in California, Ireland, France, and Massachusetts since then. He is currently a Distinguished Professor at the Department of Food Science at the University of Massachusetts where he specializes in the areas of food design and nanotechnology. He has written numerous books, published over a thousand scientific articles, been granted several patents, and presented his work at invited talks around the world. He is currently the most highly cited author in the food and agricultural sciences. He has received awards from numerous scientific organizations in recognition of his achievements. His research has been funded by the United States Department of Agriculture, National Science Foundation, NASA, and the food industry.

https://doi.org/10.1515/9783110788457-204

Chapter 1
Introduction

1.1 Introduction to nanomaterials in foods

Nanotechnology involves the design, fabrication, and application of materials with critical dimensions on the nanometer scale, which is usually taken to range from about 10 to 100 nm (i.e., 10^{-8} to 10^{-7} m), but is sometimes extended from about 1 to 1000 nm (10^{-9} to 10^{-6} m) (Cao and Wang 2010, Bertino 2022). Nanomaterials may be constructed from inorganic or organic matter, such as metals, minerals, proteins, lipids, or carbohydrates. The physicochemical and functional attributes of nanomaterials are often appreciably different from those of conventional materials because of their smaller dimensions, higher specific surface areas, and altered surface reactivities (Nile et al. 2020). Consequently, nanotechnology is being utilized to produce advanced materials with innovative or enhanced functional properties. Nanomaterials are already used in several industries because of these unique attributes, including for microelectronics, aerospace, construction, chemical, consumer goods, personal care, health care, and pharmaceutical industries (Bertino 2022). The purpose of this book is to highlight the application of nanomaterials within the modern food and agricultural industries (Figure 1.1). In particular, the ability of nanomaterials to improve the safety, quality, healthiness, and sustainability of the food supply

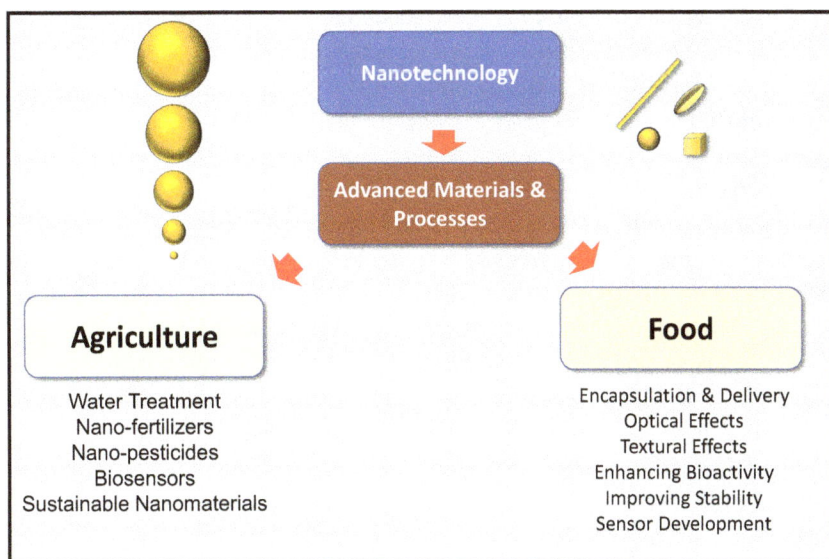

Figure 1.1: Examples of the application of nanotechnology to create Advanced materials and processes in the food and agricultural industries.

https://doi.org/10.1515/9783110788457-001

is highlighted. In this chapter, the historical development of nanotechnology is outlined, and then kinds of nanomaterials used in foods are described, including a discussion of the unique attributes of these nanomaterials. In later chapters, the application of nanotechnology in different areas of the food and agricultural industries is highlighted.

1.2 A brief history of nanotechnology

Nanotechnology was only recognized as a unique discipline in the 1980s due to the pioneering work of several scientists and engineers around the globe (McClements 2019, Bayda et al. 2020). However, nanomaterials have been created and utilized by the human race for much of its history. Some inks, such as those used by the Ancient Egyptians, consisted of carbon nanoparticles (soot) stabilized by biopolymers (like gum arabic) dispersed in water. Some types of pottery created by the Ancient Romans had glittering surfaces because they contained nanoparticle (silver or gold) coatings that scattered light into characteristic patterns. At that time, the people who created these inks and coatings did not realize they contained tiny particles, but they were still using a form of nanotechnology to create these products.

From the nineteenth century, several scientists have actively studied the properties and behavior of small particles (Bayda et al. 2020). Indeed, several famous scientists have worked in this field and produced knowledge that has become the basis for much of modern nanotechnology. In 1857, Michael Faraday prepared colloidal suspensions consisting of gold nanoparticles suspended in water, which exhibited unique optical and electronic properties. For instance, they appeared different colors depending on the particle size and light source due to their light scattering properties. In the mid-nineteenth century, Sir George Gabriel Stokes also derived an equation (known as Stokes' law) that is widely used to predict the speed at which nanoparticles rise or sink due to the forces of gravity when they are dispersed in a fluid. This equation is still widely used in the food industry to predict the stability of food products containing small particles to separation, including milks, creams, and soft drinks (McClements 2015a). In the late nineteenth century, Lord John William Strutt (Lord Rayleigh) developed an equation to describe the scattering of light waves by nanoparticle suspensions, which can be used to predict their turbidity or opacity. This equation is also still used in the food industry to predict the optical properties of foods containing small particles, such as soft drinks (McClements 2015a).

At the beginning of the twentieth century, while writing and publishing on relativity and quantum theory, Albert Einstein published important papers on the rheology of particle suspensions. Food scientists still use the equation Einstein derived to relate the relative viscosity of a dilute nanoparticle suspension to the volume fraction (ϕ) of particles it contains: $\eta_R = 1 + 2.5\phi$ (McClements 2015a). The Einstein equation is used to predict how effective a small particle or polymer is at increasing the shear viscosity of a fluid. Even though they carried out research that was important to understanding the properties of very small particles, none of these scientists thought of themselves as nanotechnologists, since this discipline had not been defined yet.

The development of nanotechnology as a widely recognized scientific discipline did not occur until the mid- to late-twentieth century. Richard Feynman, who later became a Nobel Prize winner for his research in quantum mechanics, is considered to have first introduced the concept of nanotechnology in his seminal lecture "There's Plenty of Room at the Bottom," which was given at an American Physical Society meeting in 1959 (Bayda et al. 2020). Feynman discussed the concept of fabricating a new generation of materials with innovative functional attributes by manipulating matter at the atomic and molecular scale. As part of this lecture, he presented the idea of producing extremely powerful microscopes, miniature electronic circuits, and tiny machines, many of which have been successfully produced and used since that time. At the time, the ideas Feynman introduced in his seminal lecture were not widely known, and it was only decades afterward that it became seen as one of the most pivotal moments in nanotechnology history.

The first instance of the expression "nano-technology" being used in the scientific literature was by Professor Norio Taniguchi from the Tokyo Science University in 1974 (Bayda et al. 2020). He used this expression in a manuscript to describe precision manufacturing processes that could manipulate matter at the atomic or molecular scales to create materials with innovative properties. However, it was not until the 1980s that nanotechnology became a mainstream discipline, which was largely due to the pioneering work of Eric Drexler, an American engineer working at the Massachusetts Institute of Technology during that period (Bayda et al. 2020). He published several influential scientific articles and books on nanotechnology. His popular science book *Engines of Creation: The Coming Era of Nanotechnology* (1986) described potential beneficial and adverse effects of using nanotechnology. Specifically, he discussed the creation of tiny machines capable of manipulating atoms and molecules to create new devices that might be used in computers, medicine, and material science. In practice, there are few examples of the tiny machines envisioned by Drexler that are commercially viable today. However, there are many examples of more prosaic uses of nanotechnology to create advanced materials with a wide range of industrial applications. Carbon nanotubes have been used to fabricate very strong, stiff, and lightweight materials that can be used to construct consumer products, like tennis rackets, baseball bats, golf clubs, bicycle frames, cars, boats, and airplanes. These materials are assembled from hollow tubes consisting of multiple carbon atoms covalently bonded to each other. Metal nanoparticles, such as those fabricated from silver, have been used as antimicrobials in packaging materials and clothes. Inorganic nanoparticles, like those fabricated from titanium dioxide, have been utilized in cosmetics and sunscreens due to their ability to strongly scatter light waves, thereby helping to protect the skin from

UV-radiation damage. The performance of electronic goods, such as computers and phones, has been enhanced by using nanoscale components to construct circuits and memory chips, which has led to reductions in their physical dimensions and energy expenditure. In the pharmaceutical industry, organic and inorganic nanoparticles have been utilized to encapsulate drugs and then control or target their release inside the human body (Ragelle et al. 2017).

The use of nanotechnology to create innovative materials has already been shown to be commercially successful in many industries. It is highly likely that its range of applications will continue to expand in the future as scientists make new discoveries on how to design, fabricate, characterize, and utilize nanomaterials.

1.3 Nanotechnology in foods

It is sometimes assumed that nanomaterials have only been used in foods for the past couple of decades or so. However, nanoparticles are naturally present in many commonly consumed foods. For instance, both cow's and human milk naturally contain small protein nanoparticles (casein micelles), that are complex assemblies of proteins, calcium, and phosphorous (Holt et al. 2013, Sadiq et al. 2021). These nanoparticles increase the water-dispersibility, digestibility, and absorption of these valuable nutrients, which are crucial for growing infants. Many oil seeds, including soybeans and nuts, contain natural oil bodies that have dimensions in the nanoscale range (Zaaboul et al. 2022). Consequently, humans have been consuming nanoparticles for millennia, albeit without their knowledge. More recently, commonly used processing technologies, like high-pressure valve homogenization, have been used to produce homogenized cow's milk or soft drinks that contain oil droplets with dimensions in the nanoscale range (McClements 2015a). The size of the oil droplets in these products is usually reduced into the nanoscale to increase their resistance to creaming during storage or to control the optical properties of the beverage. Before the term "nanoparticle" was coined, many researchers used the term "colloid" to refer to the tiny particles in foods and other products. By definition, a food colloid has dimensions ranging from around 1 to 1000 nm (Dickinson 1992), which is the same range often meant by the term 'nanoparticle' now. Thus, many researchers have worked on nanoparticles over the past several decades, without using this term. Nevertheless, the results of this research can still be used to inform the discipline of food nanotechnology.

The growing interest in the application of nanotechnology in foods is clearly shown by the rapid rise in the number of published scientific articles containing the keywords "food" and "nanotechnology" (Figure 1.2). This number increases greatly when other related terms are included, such as "nanoparticle," "nanoemulsion," or "nanofiber." Following the first publication in 2002, there have been more than 7,300 additional papers, with over 195,000 citing articles. Both the number of articles and

Food + Nanotechnology

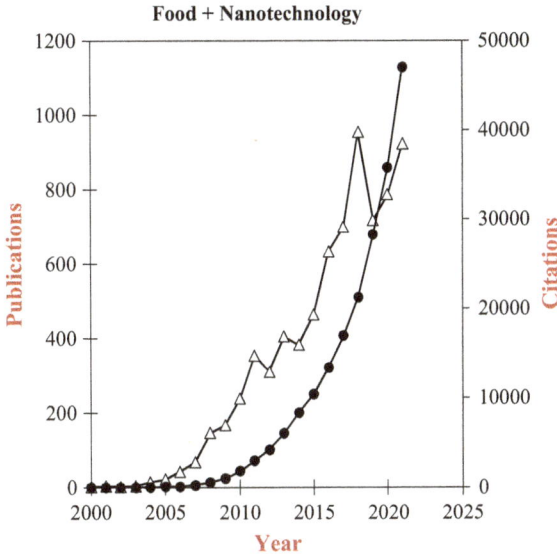

Figure 1.2: The number of publications and citations using the terms "food" and "nanotechnology" is increasing over time, showing the great interest in this area.

citations per year have tended to rise over time, highlighting this is still a highly active research area. It should be noted, however, that many food companies are reluctant to use the term "nano-" to describe their products (even when they do contain nanoparticles) because of a potentially negative consumer perception around this term.

In later chapters in this book, a number of potential applications of nanotechnology to improve the food and agriculture system are discussed, including their use to modify food quality, safety, and nutrition, as well as to create nanopesticides and nanofertilizers, innovative packaging materials, and smart sensors.

1.4 Nature of food nanomaterials

Foods contain many kinds of objects that are too small to observe with the unaided human eye, such as the air bubbles in whipped cream, the fat globules in milk, the fat crystals in butter and margarine, the plant cells in fruits and vegetables, the ice crystals in ice cream, and the starch granules in sauces and desserts (Figure 1.3). Nevertheless, most of these small objects are too large to be considered as nanomaterials; i.e., they have dimensions exceeding 100 nm (or even 1000 nm). Even so, there are smaller objects in foods that can be considered to be nanomaterials because they do have dimensions in the 1 to 100 nm or 1 to 1000 nm range, depending on the definition of nanotechnology used, such as the small oil droplets found in soft drinks and homogenized milk, the small oil bodies found in soybeans and

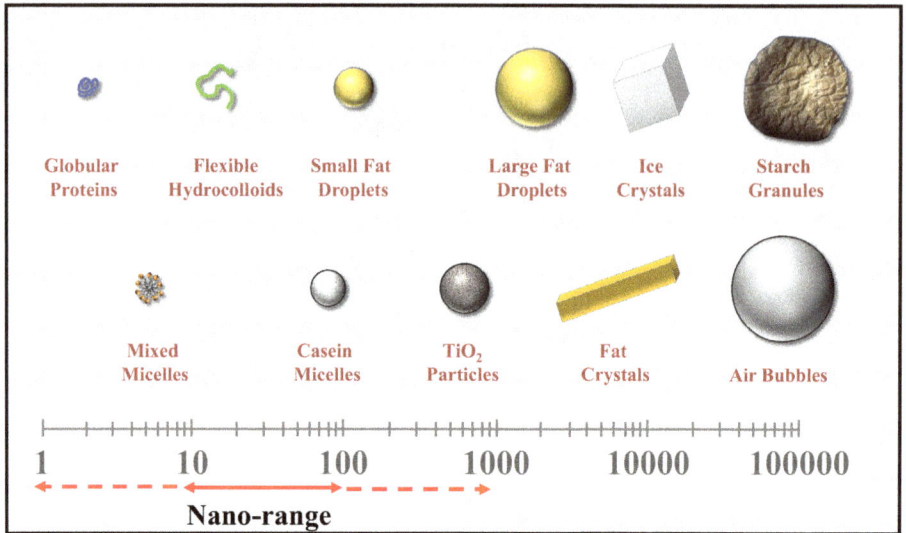

Figure 1.3: Representative sizes of particles found in foods and beverages. The nanoscale is usually taken to range from around 10 to 100 nm but is sometimes taken to range from 1 to 1000 nm.

nuts, the nanofibers used as texturing agents in some foods, and the nanoparticles found in many food-grade colloidal delivery systems. These nanomaterials may have different morphologies depending on their origin and processing, such as nanoparticles, nanofibers, nanotubes, and nano-porous materials. Nanomaterials may be naturally present within foods, they may be intentionally added, or they may be unintentionally created during food processing (McClements 2019).

Examples of naturally occurring nanoparticles in foods are the casein micelles in milk, the oil bodies in soybeans, and the lipoproteins in eggs (Holt et al. 2013, Naderi et al. 2017, Sadiq et al. 2021, Zaaboul et al. 2022). Engineered nanoparticles may be created using a variety of chemical, physical, and biological approaches. For instance, lipid nanoparticles can be formed by homogenizing oil, water, and a suitable emulsifier at high pressure using a microfluidizer (McClements 2015b). These nano-ingredients are often utilized to enhance the performance of food and beverage products, e.g., to increase the optical clarity of soft drinks by reducing light scattering, to increase the resistance to creaming or sedimentation by reducing gravitational forces, or to increase the bioavailability of hydrophobic nutrients by increasing their digestion, solubilization, and absorption rate. In some cases, a food manufacturer may not be aware that they are creating nanoparticles in their foods. For example, the homogenization of cow's milk was carried out to improve its resistance to creaming, but this process typically leads to the formation of a large fraction of particles with diameters below about 300 nm (Dickinson 1992). There is currently a great deal of research on the utilization of nanotechnology to design and fabricate engineered nanoparticles to improve the appearance, texture, stability, flavor, and nutrition of foods. For instance, edible nanoparticles are being developed to encapsulate omega-3 oils, vitamins, nutraceuticals, flavors, colors, and preservatives (McClements 2015b, McClements 2020).

Several kinds of common food processing operations can lead to the generation of nanomaterials within foods, including grinding, homogenization, and cooking (Gupta et al. 2016, Fellows 2017). In some cases, the manufacturer intended to produce the nanoparticles to obtain some desirable effect. In other cases, the manufacturer might not even be aware that they have created nanoparticles in their foods. Nanomaterials can also inadvertently get into foods *via* other routes. For instance, nanoparticles contained within packaging materials may migrate into any foods they are in contact with. Alternatively, foods may be contaminated with nanoparticles in their environment, e.g., nano-enabled pesticides or fertilizers used to treat agricultural crops or airborne nanoparticles (such as soot or smog). It is often useful to know the types and concentrations of nanoparticles in foods and how they impact their properties.

The characteristics of food nanoparticles influence their impact on the physicochemical and functional attributes of foods, as well as their potential toxicity. Nanoparticles vary in their composition, size, shape, aggregation state, physical state, surface chemistry, and charge, which impacts their behavior in foods and the human body (Figure 1.4). Food nanoparticles are often classified based on their compositions (organic or inorganic) and digestibility within the human gut (digestible or indigestible) because these characteristic influence their potential toxicity (McClements et al. 2017). Indigestible inorganic nanoparticles (like those comprising titanium dioxide) are usually considered to be more of a health risk than digestible organic nanoparticles (like those comprising fat, protein, or starch). Nevertheless, it is important to establish

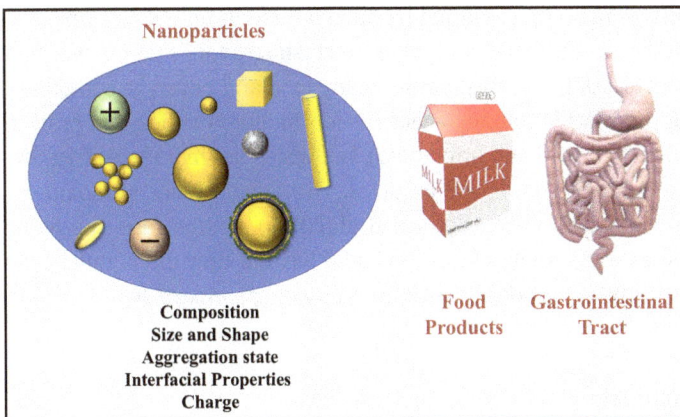

Figure 1.4: The nanoparticles used in food and agriculture may vary in their compositions, sizes, shapes, aggregation state, interfacial properties, and charges, which influences their impacts on food properties, as well as their gastrointestinal fate.

the potential toxicity of food nanoparticles on a case-by-case basis. The potential health concerns associated with nanoparticles are discussed in more detail in Chapter 6. In the remainder of this section, different kinds of nanoparticles that may be used in food and agriculture applications are highlighted.

1.4.1 Inorganic nanomaterials

Several kinds of nanoparticles used within the food and agricultural industries are fabricated from inorganic materials, including gold, silver, copper, iron, titanium, silicon, zinc, and their oxides (Pietroiusti et al. 2016). Food ingredients based on these inorganic materials are often powders that contain a broad range of different particle sizes, with some of them falling within the nanoscale range. For instance, copper hydroxide nanoparticles are present in commercial pesticides that are sprayed onto crops (Cota-Ruiz et al. 2018). Silver nanoparticles are used as antimicrobials in some packaging materials (Ahmad et al. 2021). Silicon dioxide nanoparticles are present in anticaking agents added to food powders, such as coffee creamers, hot beverages, and spices, to stop caking and increase their flowability (Younes et al. 2018b). Titanium dioxide nanoparticles are used in food products like candy, chewing gum, donuts, dressings, yogurt, and milk powders to increase their lightness or brightness (Weir et al. 2012). The presence of nanoparticles in foods often enhances their functional attributes, but there is concern about the potential adverse health effects of some kinds of inorganic nanoparticles (Choi et al. 2013, McClements and Xiao 2017). For instance, the French government recently banned the use of titanium dioxide in foods (Audran 2020). Indeed, a recent review of the safety of titanium dioxide particles by the European Food Safety Authority (EFSA) Panel on Food Additives and Flavorings concluded that genotoxicity of titanium dioxide particles could not be ruled out based on the available evidence, so they concluded that it could no longer be considered as safe for use as a food additive (Younes et al. 2021). For this reason, many food and beverage companies are reformulating their products to find replacements for titanium dioxide that provide similar optical properties (lightness and brightness), but which are more label friendly, such as calcium-, starch-, or protein-based particles. The safety of silicon dioxide in foods has also been reviewed by EFSA (Younes et al. 2018a). In this case, the panel reported no strong evidence of toxicity of this food additive, but they did highlight the need for more long-term studies, especially on silicon dioxide nanoparticles.

1.4.2 Organic nanomaterials

Many foods contain synthetic or natural organic nanoparticles that consist of carbon-based materials, like proteins, polysaccharides, lipids, or phospholipids (McClements et al. 2017). These nanoparticles may be naturally present within the food or beverage, or they may be intentionally or unintentionally created during the manufacturing process. Raw milk naturally contains casein micelles, which are nanoscale particles comprising protein, calcium, and phosphate (Rehan et al. 2019). Homogenized milk also contains these casein micelles, as well as nanoscale fat droplets created when the large milk fat globules are broken down inside homogenizers (Singh and Gallier 2017). Some kinds of plant-based milks contain fat or protein nanoparticles that are formed when the cellular structure of plants are broken down by mechanical, chemical, or enzymatic methods (McClements et al. 2019).

Many of the desirable flavors added to foods and beverages are insoluble in water and are therefore converted into nanoparticles prior to utilization. For instance, soft drinks often contain nanoscale fat droplets that have hydrophobic flavors inside, like orange, lemon, or lime oils (Piorkowski and McClements 2014). Small fat droplets are used for this purpose because they are highly resistant to gravitational forces and therefore do not tend to cream or sediment during storage. Moreover, the small size of these fat droplets means they only scatter light weakly, so they can be used in products that need to be clear or only slightly turbid. Similarly, water-insoluble vitamins (like vitamins A, D, or E) or nutraceuticals (like β-carotene, lycopene, or curcumin) can be loaded into small fat droplets that are then incorporated into functional foods or beverages (McClements 2015c, Akhavan et al. 2018). In all these applications, the fat droplets are stabilized against aggregation by coating them with a suitable emulsifier.

Nanofibers can be isolated from many natural sources. For instance, cellulose nanofibers and nanocrystals can be extracted from wood, some plants (like cotton), and agricultural waste streams (Pennells et al. 2020, Liu et al. 2021, Nagarajan et al. 2021). These nanofibers can then be used as additives in foods and packaging materials as texture modifiers, fat replacers, or digestion modifiers. Chitin nanofibers can be extracted from crab shells, insect shells, and fungi (Tian and Liu 2020, Yang et al. 2020), which can be used for similar purposes. Typically, these polysaccharide-based nanofibers are obtained by treating the source materials with strong acids and/or enzymes to disrupt their structures. Various kinds of carbon nanomaterials can also be utilized in the food industry, such as carbon nanodots, graphene oxide, carbon nanotubes, carbon nanofibers, and fullerene (Moradi et al. 2022, Zhao et al. 2022, Chandel et al. 2022). These nanomaterials are widely used to construct innovative packaging materials and sensors for application in the food and agricultural industries.

Only a few examples of inorganic and organic nanoparticles have been given here. In later chapters, several other kinds of nanoparticles that have been or can be utilized in the food industry to create novel materials are discussed.

1.4.3 Preparation of nanoparticles

Nanoparticles can either be isolated from natural sources (like the oil bodies from oil seeds) or they can be fabricated (like the nanoscale fat droplets in soft drinks or the titanium dioxide nanoparticles in whitening agents). A diverse range of methods can be used to fabricate food-grade nanoparticles, which can be classified as top-down or bottom-up approaches (Figure 1.5) (McClements 2015b). In top-down approaches, bulk materials or larger particles are usually broken down into smaller particles through the application of intense mechanical forces generated by specialized devices (such as mills or homogenizers). For instance, solid materials (like powdered starch, nutraceuticals, or fertilizers) can be converted into nanoparticles by grinding them using a suitable milling device, like a high-energy ball mill (Moschwitzer 2013, Pohshna et al. 2020, Lu and Tian 2021). In contrast, liquid materials (like triacylglycerol, essential, or flavor oils) can be converted into nanoemulsions by homogenizing them with an aqueous phase containing a suitable emulsifier using a sonicator, homogenizer, or microfluidizer (Aswathanarayan and Vittal 2019, Choi and McClements 2020). These devices generate intense cavitational, shear, and turbulent fluid flow profiles that break up, intermingle, and reduce the size of the oil phase leading to the formation of emulsifier-coated nanoscale oil droplets. Selective chemical or enzymatic degradation methods are another example of top-down approaches. As mentioned earlier, nanocellulose and nanochitin can be obtained by treating natural materials (such as wood, cotton, or crab shells) with strong acids, often in combination with mechanical forces (Yang et al. 2020, Nagarajan et al. 2021).

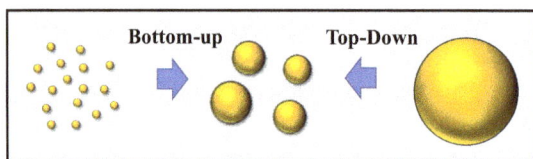

Figure 1.5: Nanomaterials can be created using bottom-up or top-down approaches. The bottom-up approach consists of assembling nanomaterials from smaller units, whereas the top-down approach involves breaking down large objects into smaller units.

In bottom-up approaches, nanoparticles are usually fabricated using methods that promote the clustering of molecules by increasing the attractive forces between them, such as antisolvent precipitation, controlled aggregation, and coacervation methods. The antisolvent precipitation approach involves dissolving a material within a suitable solvent and then injecting the resulting solution into an antisolvent (Joye and McClements 2013). After mixing, the material becomes insoluble and tends to precipitate, leading to the formation of nanoparticles under appropriate conditions. This method is often used to create nanoparticles from hydrophobic proteins, such as zein and gliadin. The proteins are dissolved within a concentrated alcohol solution (such as 80% ethanol and 20% water) and then injected into an aqueous solution. After injection, the environment around the proteins becomes more polar because the alcohol solution is diluted with water, which causes the proteins to precipitate and form nanoparticles. These nanoparticles can be stabilized against aggregation by adding emulsifiers or biopolymers that adsorb to their surfaces and protect them. They can also be loaded with hydrophobic bioactive agents by dissolving the bioactive agents in the protein-ethanol solution before adding to water. Protein nanoparticles typically have dimensions of a few hundred nanometers, but smaller ones can be created under appropriate conditions. This approach may be useful for applications in food and beverage products that are expected to be optically clear since nanoparticles less than about 50 nm do not scatter light strongly.

Lipophilic nanocrystals can also be formed using the antisolvent precipitation approach. For instance, nutraceutical nanocrystals (like curcumin) have been formed by dissolving them in a concentrated alcohol solution and then adding them to water (Zou et al. 2016). Alternatively, some nutraceuticals can be added to a concentrated alkaline solution (where they are soluble), which is then injected into an acidic or neutral solution (where they are insoluble), which leads to the formation of nutraceutical nanocrystals (Dai et al. 2019).

Nanocrystals or bioactive-loaded nanoparticles can be utilized to fortify food and beverage products with nutrients and nutraceuticals, like omega-3 fatty acids, vitamins, curcumin, carotenoids, and polyphenols. Encapsulation of these bioactives can improve their handling, water-dispersibility, food matrix compatibility, stability, flavor profile, and bioavailability.

Bottom-up approaches are also utilized to fabricate many kinds of metal-based nanoparticles, such as those prepared from gold, silver, or copper. For example, silver nanoparticles can be formed by adding a reducing agent (such as ethylene glycol) to a solution containing soluble metal ions (such as silver nitrate) in the presence of a suitable blocking or capping agent that prevents particle aggregation (such as polyvinyl pyrrolidone) (Zhang et al. 2018). The methods traditionally utilized to fabricate metal nanoparticles often involve the use of synthetic chemicals and harsh preparation conditions. There has therefore been interest in utilizing "green chemistry" methods to fabricate this kind of nanoparticle, where synthetic chemicals are replaced by natural alternatives, such as tea, coffee, banana, or wine extracts (Villasenor and Rios 2018, Rozhin et al. 2021). In addition, some kinds of microorganisms, including bacteria, yeasts, and viruses, can be used to fabricate metal nanoparticles from natural materials. The utilization of these production methods may lead to a more sustainable and environmentally friendly approach to manufacturing food-grade nanoparticles.

The composition, size, shape, and surface chemistry of nanoparticles can be manipulated by using different ingredients and preparation methods, which enables their functional properties to be tailored to specific applications (Figure 1.4).

The design and fabrication of food-grade nanoparticles has been a major research focus in recent years, and there is a growing number of approaches being developed to produce nanoparticles for different purposes.

1.5 Unique properties of nanomaterials

There has been great interest in the application of nanomaterials within the food and other industries because they have some unique characteristics associated with their small dimensions. In general, nanomaterials may vary in their compositions, sizes, shapes, interfacial properties, physical states, and aggregation states, which means that their physicochemical and functional attributes can be controlled (Figure 1.4). In this section, a brief overview of some of the unique features of nanomaterials is given. A more detailed discussion of nanomaterial properties and characterization is given in Chapter 2.

Small size: The small dimensions of nanoparticles mean they can behave in ways that are not possible for larger particles or bulk materials. For instance, small particles can often penetrate through biological barriers, such as the biofilms formed by bacteria, the mucus layer that coats the gastrointestinal tract, or plant, animal, or microbial cell walls, much easier than larger particles. Consequently, they may have different biological effects, such as increased bioavailability and bioactivity, which could have desirable or undesirable effects on human and environmental health. As an example, nano-sized fertilizers can penetrate through the pores in plant surfaces more easily, thereby increasing the concentration of nutrients that plants absorb, which promotes growth and resilience (Dimkpa and Bindraban 2017). Similarly, antimicrobial nanoparticles may penetrate through microbial cell walls more easily than larger particles, thereby improving their ability to kill pathogenic or spoilage organisms (Wang et al. 2017). Converting micronutrients into nanoparticles can increase their bioavailability by enhancing their ability to penetrate through the mucus layer and epithelium cells (Braithwaite et al. 2014).

The small size of nanomaterials reduces the magnitude of the gravitational forces acting upon them, thereby improving their resistance to creaming or sedimentation (McClements 2015b). In addition, it also reduces the magnitude of the attractive colloidal interactions acting between them, thereby improving their resistance to aggregation. As a result, reducing the dimensions of food-grade particles into the nano-range can increase the shelf life of foods, leading to improvements in food quality and reductions in food waste.

The size of the particles in foods also influences their appearance since it impacts their interactions with light waves (McClements 2015b). When the diameter of particles is less much less than the wavelength of light (i.e., $d < 50$ nm), they only scatter light waves weakly, which means that optically clear particle dispersions can be created. This is advantageous when a food manufacturer wants to incorporate hydrophobic substances (such as oil-soluble colors, flavors, preservatives, or nutrients) into transparent foods or beverages (such as soft drinks, fortified waters, jellies, or gummies). The small size of nanoparticles also means that the average distance between each particle is reduced. This phenomenon can be utilized to create desirable textural attributes in foods. For instance, a dispersion of nanoparticles often has a higher viscosity or gel strength than a similar dispersion of larger particles because of increased particle-particle interactions associated with the fact that the particles are closer together. Thus, it may be possible to reduce the fat content of foods, without altering their desirable textural attributes, simply by reducing the particle size.

Table 1.1: Impact of particle diameter (D) on the number (N), surface area (S), and percentage of molecules at the particle surface (%S) for 1 kg of material.

D	N (kg^{-1})	S (m^2 kg^{-1})	%S
1 nm	2.08×10^{24}	6.52×10^6	100
10 nm	2.08×10^{21}	6.52×10^5	27.1
100 nm	2.08×10^{18}	6.52×10^4	3.0
1 μm	2.08×10^{15}	6.52×10^3	0.30
10 μm	2.08×10^{12}	6.52×10^2	0.030
100 μm	2.08×10^9	6.52×10^1	0.0030
1 mm	2.08×10^6	6.52	0.0003

Large surface area: The specific surface area of a particle is inversely related to its dimensions: the smaller the size, the greater the surface area. As an example, 1 kg of material in a spherical shape has a surface area of about 510 cm^2, but when it is divided into nanoparticles (100 nm) it has a surface area or about 6500 m^2 (Table 1.1). The relatively large surface area of nanomaterials is important for understanding their behavior within foods, as well as for certain applications within the food industry. For instance, the rate of lipid digestion within the gastrointestinal tract tends to increase as the droplet size decreases because of the increase in oil-water surface area available for the lipase molecules to bind to (Tan et al. 2020). The ability of nano-filters to remove contaminants from wastewater is often better than conventional filters because the former have a larger surface area for the contaminants to adsorb to (Sun et al. 2021). The sensitivity of sensors can be increased by reducing their size into the nano-range because then there is a larger surface area for the target molecules to bind to (Sargazi et al. 2022).

Higher surface reactivity: The physical properties and chemical reactivity of the molecules at a surface of a particle are usually different from those in the interior because they are surrounded by different kinds of molecules (Figure 1.6). As a result, the overall properties of a material tend to change as the particle size is decreased because then a greater fraction of the molecules are present at the surface (Table 1.1). This phenomenon may alter the chemical reactivity of nanomaterials. In addition, the properties of

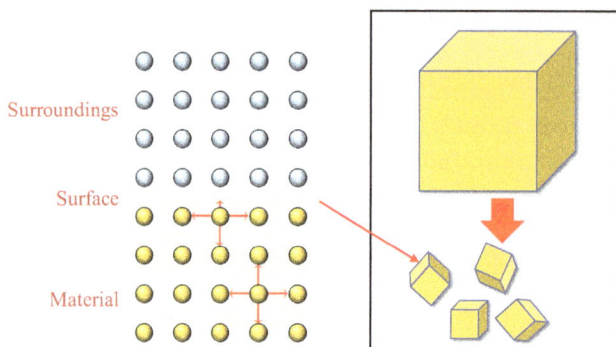

Figure 1.6: Different molecular interactions occur at the surface of particle than in its interior, which means that the surface material has different physicochemical properties than the bulk material.

the matter within particles can change substantially when they become very small as a result of *quantum effects* (Hornyak et al. 2008, Bertino 2022). These quantum effects can lead to alterations in the optical, electronic, magnetic, physical, and chemical properties of materials. Consequently, nanomaterials may behave differently than bulk materials. For instance, a macroscopic piece of gold has a yellowish appearance, but a dispersion of gold nanoparticles appears reddish or purplish depending on the particle size because the optical properties are altered. Changes in the physicochemical properties of substances when they are made very small can be utilized to produce materials with novel optical or magnetic properties (which may be useful for developing nano-enabled sensors), or to create more efficient catalysts (which may be useful for controlling chemical reactions). Nevertheless, because nanomaterials behave differently from macroscopic ones, it is important to understand any potential adverse health effects that might arise from reducing the size of materials (Chapter 6).

1.6 Conclusions

Researchers are exploring the potential of using nanomaterials for a wide range of applications within the food and agricultural industries because of their potential benefits over conventional materials. For instance, the small size, high surface area, and altered surface reactivity of nanomaterials can provide novel physicochemical properties and functional attributes. In principle, these innovative materials can therefore be used to increase the safety, healthiness, quality, and shelf life of foods. Nevertheless, it is important to ensure they do not have any adverse effects on human health or the environment. In the remainder of this book, we will examine the unique characteristics of food nanoparticles in more detail, and then highlight their potential applications within agriculture and foods. Finally, the potential toxicity of food nanoparticles is discussed, as this must be carefully considered when introducing them into the food supply chain.

References

Ahmad, S. S., O. Yousuf, R. Ul Islam and K. Younis (2021). "Silver nanoparticles as an active packaging ingredient and its toxicity". Packaging Technology and Science **34**(11–12): 653–663.

Akhavan, S., E. Assadpour, I. Katouzian and S. M. Jafari (2018). "Lipid nano scale cargos for the protection and delivery of food bioactive ingredients and nutraceuticals". Trends in Food Science & Technology **74**: 132–146.

Aswathanarayan, J. B. and R. R. Vittal (2019). "Nanoemulsions and their potential applications in food industry" Frontiers in Sustainable Food Systems 3: 1–21. doi.org/10.3389/fsufs.2019.00095

Audran, X. (2020). France: France Bans Titanium Dioxide in Food Products by January 2020. Washington, D.C., USDA: 1–3.

Bayda, S., M. Adeel, T. Tuccinardi, M. Cordani and F. Rizzolio (2020). "The history of nanoscience and nanotechnology: From chemical-physical applications to nanomedicine" Molecules **25**(1): 1–15.

Bertino, M. F. (2022). Introduction To Nanotechnology. Hackensack, NJ, World Scientific.

Braithwaite, M. C., C. Tyagi, L. K. Tomar, P. Kumar, Y. E. Choonara and V. Pillay (2014). "Nutraceutical-based therapeutics and formulation strategies augmenting their efficiency to complement modern medicine: An overview". Journal of Functional Foods 6: 82–99.

Cao, G. and Y. Wang (2010). Nanostructures and Nanomaterials: Synthesis, Properties, and Applications. Hackensack, NJ, World Scientific Series.

Chandel, M., K. Kaur, B. K. Sahu, S. Sharma, R. Panneerselvam and V. Shanmugam (2022). "Promise of nano-carbon to the next generation sustainable agriculture". Carbon **188**: 461–481.

Choi, S. J., J. K. Lee, J. Jeong and J. H. Choy (2013). "Toxicity evaluation of inorganic nanoparticles: Considerations and challenges". Molecular & Cellular Toxicology **9**(3): 205–210.

Choi, S. J. and D. J. McClements (2020). "Nanoemulsions as delivery systems for lipophilic nutraceuticals: Strategies for improving their formulation, stability, functionality and bioavailability". Food Science and Biotechnology **29**(2): 149–168.

Cota-Ruiz, K., J. A. Hernandez-Viezcas, A. Varela-Ramirez, C. Valdes, J. A. Nunez-Gastelum, A. Martinez-Martinez, M. Delgado-Rios, J. R. Peralta-Videa and J. L. Gardea-Torresdey (2018). "Toxicity of copper hydroxide nanoparticles, bulk copper hydroxide, and ionic copper to alfalfa plants: A spectroscopic and gene expression study". Environmental Pollution **243**: 703–712.

Dai, L., H. L. Zhou, Y. Wei, Y. X. Gao and D. J. McClements (2019). "Curcumin encapsulation in zein-rhamnolipid composite nanoparticles using a pH-driven method". Food Hydrocolloids **93**: 342–350.

Dickinson, E. (1992). An Introduction to Food Colloids. Oxford, UK, Oxford University Press.

Dimkpa, C. O. and P. S. Bindraban (2017). "Nanofertilizers: New products for the industry?". Journal of Agricultural and Food Chemistry. 66(26), 6462–6473.

Fellows, P. J. (2017). Food Processing Technology. Cambridge, MA, USA, Woodhead Publishing.

Gupta, A., H. B. Eral, T. A. Hatton and P. S. Doyle (2016). "Nanoemulsions: Formation, properties and applications". Soft Matter **12**(11): 2826–2841.

Holt, C., J. A. Carver, H. Ecroyd and D. C. Thorn (2013). "Invited review: Caseins and the casein micelle: Their biological functions, structures, and behavior in foods". Journal of Dairy Science **96**(10): 6127–6146.

Hornyak, G. L., J. Dutta, H. F. Tibbals and A. K. Rao (2008). Introduction to Nanoscience. Boca Raton, FL, CRC Press.

Joye, I. J. and D. J. McClements (2013). "Production of nanoparticles by anti-solvent precipitation for use in food systems". Trends in Food Science & Technology **34**(2): 109–123.

Liu, Y. W., S. Ahmed, D. E. Sameen, Y. Wang, R. Lu, J. W. Dai, S. Q. Li and W. Qin (2021). "A review of cellulose and its derivatives in biopolymer-based for food packaging application". Trends in Food Science & Technology **112**: 532–546.

Lu, H. and Y. Q. Tian (2021). "Nanostarch: Preparation, modification, and application in pickering emulsions". Journal of Agricultural and Food Chemistry **69**(25): 6929–6942.

McClements, D. J. (2015a). Food Emulsions: Principles, Practice and Techniques. Boca Raton, CRC Press.

McClements, D. J. (2015b). Nanoparticle- and Microparticle-based Delivery Systems. Boca Raton, FL, CRC Press.

McClements, D. J. (2015c). "Nanoscale nutrient delivery systems for food applications: improving bioactive dispersibility, stability, and bioavailability". Journal of Food Science **80**(7): N1602–N1611.

McClements, D. J. (2019). Future Foods: How Modern Science Is Transforming the Way We Eat. Cham, Switzerland, Springer Scientific.

McClements, D. J. (2020). "Nano-enabled personalized nutrition: Developing multicomponent-bioactive colloidal delivery systems" Advances in Colloid and Interface Science **282**: 1–15.

McClements, D. J., E. Newman and I. F. McClements (2019). "Plant-based milks: A review of the science underpinning their design, fabrication, and performance". Comprehensive Reviews in Food Science and Food Safety **18**(6): 2047–2067.

McClements, D. J. and H. Xiao (2017). "Is nano safe in foods? Establishing the factors impacting the gastrointestinal fate and toxicity of organic and inorganic food-grade nanoparticles" Npj Science of Food **1**(6): 1–13.

McClements, D. J., H. Xiao and P. Demokritou (2017). "Physicochemical and colloidal aspects of food matrix effects on gastrointestinal fate of ingested inorganic nanoparticles". Advances in Colloid and Interface Science **246**: 165–180.

Moradi, M., R. Molaei, S. A. Kousheh, J. T. Guimaraes and D. J. McClements (2022)."Carbon dots synthesized from microorganisms and food by-products: Active and smart food packaging applications". Critical Reviews in Food Science and Nutrition. DOI:10.1080/10408398.2021.2015283

Moschwitzer, J. P. (2013). "Drug nanocrystals in the commercial pharmaceutical development process". International Journal of Pharmaceutics **453**(1): 142–156.

Naderi, N., J. D. House, Y. Pouliot and A. Doyen (2017). "Effects of high hydrostatic pressure processing on hen egg compounds and egg products". Comprehensive Reviews in Food Science and Food Safety **16**(4): 707–720.

Nagarajan, K. J., N. R. Ramanujam, M. R. Sanjay, S. Siengchin, B. S. Rajan, K. S. Basha, P. Madhu and G. R. Raghav (2021). "A comprehensive review on cellulose nanocrystals and cellulose nanofibers: Pretreatment, preparation, and characterization". Polymer Composites **42**(4): 1588–1630.

Nile, S. H., V. Baskar, D. Selvaraj, A. Nile, J. B. Xiao and G. Y. Kai (2020). "Nanotechnologies in Food Science: Applications, Recent Trends, and Future Perspectives" Nano-Micro Letters **12**(45), 1–35.

Pennells, J., I. D. Godwin, N. Amiralian and D. J. Martin (2020). "Trends in the production of cellulose nanofibers from non-wood sources". Cellulose **27**(2): 575–593.

Pietroiusti, A., A. Magrini and L. Campagnolo (2016). "New frontiers in nanotoxicology: Gut microbiota/microbiome-mediated effects of engineered nanomaterials". Toxicology and Applied Pharmacology **299**: 90–95.

Piorkowski, D. T. and D. J. McClements (2014). "Beverage emulsions: Recent developments in formulation, production, and applications". Food Hydrocolloids **42**: 5–41.

Pohshna, C., D. R. Mailapalli and T. Laha (2020). Synthesis of Nanofertilizers by Planetary Ball Milling. Sustainable Agriculture Reviews 40. E. Lichtfouse. **40**: 75–112.

Ragelle, H., F. Danhier, V. Preat, R. Langer and D. G. Anderson (2017). "Nanoparticle-based drug delivery systems: A commercial and regulatory outlook as the field matures". Expert Opinion on Drug Delivery **14**(7): 851–864.

Rehan, F., N. Ahemad and M. Gupta (2019). "Casein nanomicelle as an emerging biomaterial-A comprehensive review". Colloids and Surfaces B-Biointerfaces **179**: 280–292.

Rozhin, A., S. Batasheva, M. Kruychkova, Y. Cherednichenko, E. Rozhina and R. Fakhrullin (2021). "Biogenic silver nanoparticles: synthesis and application as antibacterial and antifungal agents" Micromachines **12**(12): 1–17.

Sadiq, U., H. Gill and J. Chandrapala (2021). "Casein micelles as an emerging delivery system for bioactive food components" Foods **10**(8): 1–30.

Sargazi, S., I. Fatima, M. H. Kiani, V. Mohammadzadeh, R. Arshad, M. Bilal, A. Rahdar, A. M. Diez-Pascual and R. Behzadmehr (2022). "Fluorescent-based nanosensors for selective detection of a wide range of biological macromolecules: A comprehensive review". International Journal of Biological Macromolecules **206**: 115–147.

Singh, H. and S. Gallier (2017). "Nature's complex emulsion: The fat globules of milk". Food Hydrocolloids **68**: 81–89.

Sun, H. Z., Y. L. Zou, H. Y. Kaw, L. Y. Wang, G. Wang, J. L. Zhou, L. Y. Meng and D. H. Li (2021). "Carbon nanofibers-based nanoconfined liquid phase filtration for the rapid removal of chlorinated pesticides from ginseng extracts". Journal of Agricultural and Food Chemistry **69**(32): 9434–9442.

Tan, Y., Z. Zhang, J. Liu, H. Xiao and D. J. McClements (2020). "Factors impacting lipid digestion and nutraceutical bioaccessibility assessed by standardized gastrointestinal model (INFOGEST): Oil droplet size". Food & Function **11**: 9936–9946.

Tian, B. R. and Y. M. Liu (2020). "Chitosan-based biomaterials: From discovery to food application". Polymers for Advanced Technologies **31**(11): 2408–2421.

Villasenor, M. J. and A. Rios (2018). "Nanomaterials for water cleaning and desalination, energy production, disinfection, agriculture and green chemistry". Environmental Chemistry Letters **16**(1): 11–34.

Wang, L. L., C. Hu and L. Q. Shao (2017). "The antimicrobial activity of nanoparticles: Present situation and prospects for the future". International Journal of Nanomedicine **12**: 1227–1249.

Weir, A., P. Westerhoff, L. Fabricius, K. Hristovski and N. von Goetz (2012). "Titanium dioxide nanoparticles in food and personal care products". Environmental Science & Technology **46**(4): 2242–2250.

Yang, X. F., J. Liu, Y. Pei, X. J. Zheng and K. Y. Tang (2020). "Recent progress in preparation and application of nano-chitin materials". Energy & Environmental Materials **3**(4): 492–515.

Younes, M., P. Aggett, F. Aguilar, R. Crebelli, B. Dusemund, M. Filipic, M. J. Frutos, P. Galtier, D. Gott, U. Gundert-Remy, G. G. Kuhnle, J. C. Leblanc, I. T. Lillegaard, P. Moldeus, A. Mortensen, A. Oskarsson, I. Stankovic, I. Waalkens-Berendsen, R. A. Woutersen, M. Wright, P. Boon, D. Chrysafidis, R. Gurtler, P. Mosesso, D. Parent-Massin, P. Tobback,

N. Kovalkovicova, A. M. Rincon, A. Tard and C. Lambre (2018a). "Re-evaluation of silicon dioxide (E 551) as a food additive". EFSA Journal **16**(1): e05088.

Younes, M., P. Aggett, F. Aguilar, R. Crebelli, B. Dusemund, M. Filipic, M. J. Frutos, P. Galtier, D. Gott, U. Gundert-Remy, G. G. Kuhnle, J. C. Leblanc, I. T. Lillegaard, P. Moldeus, A. Mortensen, A. Oskarsson, I. Stankovic, I. Waalkens-Berendsen, R. A. Woutersen, M. Wright, P. Boon, D. Chrysafidis, R. Gurtler, P. Mosesso, D. Parent-Massin, P. Tobback, N. Kovalkovicova, A. M. Rincon, A. Tard, C. Lambre and E. P. F. A. Nutrien (2018b). "Re-evaluation of silicon dioxide (E 551) as a food additive" Efsa Journal **16**(1): 1–70.

Younes, M., G. Aquilina, L. Castle, K. H. Engel, P. Fowler, M. J. F. Fernandez, P. Furst, U. Gundert-Remy, R. Gurtler, T. Husoy, M. Manco, W. Mennes, P. Moldeus, S. Passamonti, R. Shah, I. Waalkens-Berendsen, D. Waffle, E. Corsini, F. Cubadda, D. De Groot, R. FitzGerald, S. Gunnare, A. C. Gutleb, J. Mast, A. Mortensen, A. Oomen, A. Piersma, V. Plichta, B. Ulbrich, H. Van Loveren, D. Benford, M. Bignami, C. Bolognesi, R. Crebelli, M. Dusinska, F. Marcon, E. Nielsen, J. Schlatter, C. Vleminckx, S. Barmaz, M. Cart, C. Civitella, A. Giarola, A. M. Rincon, R. Serafimova, C. Smeraldi, J. Tarazona, A. Tard, M. Wright and E. P. F. A. Flavouri (2021). "Safety assessment of titanium dioxide (E171) as a food additive" Efsa Journal **19**(5): 1–130.

Zaaboul, F., Q. L. Zhao, Y. J. Xu and Y. F. Liu (2022). "Soybean oil bodies: A review on composition, properties, food applications, and future research aspects" Food Hydrocolloids **124**. doi.org/10.1016/j.foodhyd.2021.107296.

Zhang, Z., W. F. Shen, J. Xue, Y. M. Liu, Y. W. Liu, P. P. Yan, J. X. Liu and J. G. Tang (2018). "Recent advances in synthetic methods and applications of silver nanostructures" Nanoscale Research Letters **13**: 1–18.

Zhao, L. L., M. Zhang, A. S. Mujumdar and H. X. Wang (2022). "Application of carbon dots in food preservation: A critical review for packaging enhancers and food preservatives". Critical Reviews in Food Science and Nutrition. DOI: 10.1080/10408398.2022.2039896

Zou, L. Q., B. J. Zheng, R. J. Zhang, Z. P. Zhang, W. Liu, C. M. Liu, H. Xiao and D. J. McClements (2016). "Enhancing the bioaccessibility of hydrophobic bioactive agents using mixed colloidal dispersions: Curcumin-loaded zein nanoparticles plus digestible lipid nanoparticles". Food Research International **81**: 74–82.

Chapter 2
Nanomaterial properties and their characterization

2.1 Introduction

The properties of nanomaterials, such as their composition, concentration, dimensions, shape, aggregation state, physical state, and electrical charge, play a critical role in determining their functional attributes and their impact on food and beverage properties (Figure 2.1). The focus of this chapter is therefore to discuss each of these particle characteristics, highlight their impact on food properties, and review methods of characterizing them. Special emphasis is given to food-grade nanoparticles because these are currently the most common nanomaterials used in foods and beverages. Examples of different kinds of edible nanoparticles are summarized in Table 2.1.

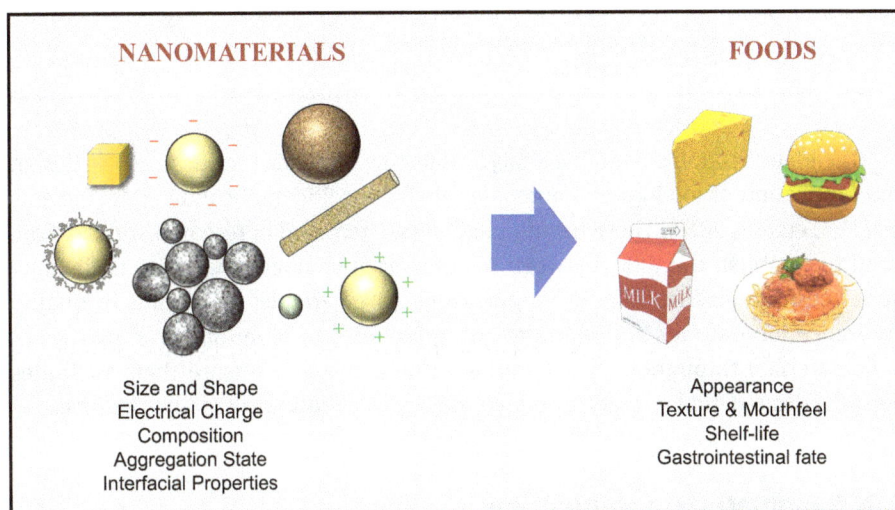

Figure 2.1: The characteristics of nanomaterials impact their effects on the properties of foods.

Table 2.1: Typical compositions, structures, and sizes of organic and inorganic nanoparticles used in foods.

Particle type	Composition	Structure	Diameter
Micelles	Surfactants, cosurfactants and/or cosolvents	Surfactant clusters	5–20 nm
Microemulsions	Surfactants, cosurfactants and/or cosolvents	Surfactant clusters with oily core	10–100 nm

https://doi.org/10.1515/9783110788457-002

Table 2.1 (continued)

Particle type	Composition	Structure	Diameter
Nanoliposomes	Phospholipids	Phospholipid bilayers	30–1000 nm
Nanoemulsions	Lipids and emulsifiers	Emulsifier-coated liquid oil droplets in water	30–1000 nm
Solid lipid nanoparticles and nanostructure lipid carriers	Lipids and emulsifiers	Emulsifier-coated solid fat particles in water	30–1000 nm
Biopolymer nanoparticles	Proteins and/or polysaccharides	Biopolymer microgels in water	30–1000 nm
Carbon nanoparticles	Carbon	Small solid particles	5–1000 nm
Metal nanoparticles	Metals or metal oxides, like gold, silver, copper, titanium, and silicon	Small solid particles	5–1000 nm

The identification of suitable analytical testing methods is critical for the design and fabrication of food-grade nanomaterials for specific applications (McClements and McClements 2016). These testing methods are required in research and development to establish the links between nanomaterial characteristics and their functional performance in different applications. They are also important in quality assurance laboratories and manufacturing facilities to monitor the properties of nanomaterials throughout the manufacturing process to ensure their attributes meet the desired quality criteria, as well as to predict how end products behave.

2.2 Nanomaterial composition

2.2.1 Importance

As mentioned in Chapter 1, the nanomaterials used in the food and agricultural industries may be constructed from various kinds of organic and/or inorganic materials (Table 2.1). The composition of nanomaterials plays an important role in determining their functional performance and so must be carefully controlled and monitored. Inorganic nanomaterials are typically assembled from metals or their oxides, such as gold, silver, copper, titanium, zinc, and silicon. Organic nanomaterials consist of those assembled from typical food ingredients, such as proteins, carbohydrates, or lipids, as well as those assembled from carbon (such as carbon nanotubes or nanodots).

Some of the ways that nanomaterial composition impacts their physicochemical properties and functional performance are highlighted below (McClements 2015b):

- *Physical properties*: The density, rheological, optical, and phase transition properties of a nanomaterial depend on its composition. These physical properties influence the stability, rheology, and appearance of foods or packaging materials that the nanomaterials are incorporated into. Consequently, it is often important to select a nanomaterial that has appropriate physical properties for specific applications.
- *Chemical reactivity*: The susceptibility of a nanomaterial to specific types of chemical reaction (such as oxidation, reduction, hydrolysis, or polymerization) depends on its composition. For instance, polyunsaturated lipids, such as omega-3 fatty acids or carotenoids, are highly susceptible to oxidation, which can limit their application in some food and beverage products. Many proteins are natural antioxidants and may therefore be incorporated into edible nanoparticles to retard the oxidation of encapsulated bioactive agents (like polyunsaturated lipids).
- *Aggregation stability*: The susceptibility of nanoparticles to aggregate in solution when exposed to different environmental conditions (such as pH, ionic strength, temperature, and additive type) depends on their composition. For instance, protein nanoparticles tend to be resistant to aggregation at pH values much higher or lower than their isoelectric points because they have a high net charge, which leads to a strong electrostatic repulsion. However, they tend to aggregate near their isoelectric point because they lose their charge, which weakens the electrostatic repulsion. In contrast, polysaccharide nanoparticles may be stable over a much wider range of pH values because they are mainly stabilized by steric repulsion.
- *Digestibility*: The digestion of a nanomaterial within the human gastrointestinal tract depends on the material it is constructed from. For example, starches are digested by amylases in the mouth and small intestine, proteins are digested by proteases in the stomach and small intestine, lipids are digested by lipases in the stomach and small intestine, and dietary fibers are fermented and digested by enzymes released by bacteria residing in the colon. In contrast, many inorganic materials are indigestible and non-fermentable under gastrointestinal conditions. Different food-grade materials can therefore be used to assemble nanomaterials that are broken down in different regions of the human gut, which may be useful for developing targeted delivery systems.
- *Economics, regulations, and labeling*: The components selected to fabricate a nanomaterial are also important for various practical reasons. For instance, the cost of these components can vary considerably, which may influence the economic viability of the final product. The components used to formulate nanomaterials should be consistent with any government regulations in the countries where they are employed. There are often restrictions or usage limits on the types of components that can be incorporated into foods or packaging materials. For example, titanium dioxide nanoparticles were recently

banned from use in France because of concerns about their toxicity. The components used to assemble any edible nanomaterials should also meet labelling expectations. For instance, they may have to be Kosher, vegetarian, vegan, all-natural, or non-allergenic. Consequently, food manufacturers must carefully consider the nature of the materials used to assemble nanomaterials utilized within food products.

2.2.2 Characterization

A variety of analytical methods can be used to provide information about the composition of the nanomaterials used in foods, including those based on chemical, chromatography, electrophoresis, mass spectrometry, and spectroscopy methods (McClements and McClements 2016). The most suitable method for a particular application depends on the nature of the nanomaterial being analyzed. In some cases, the nanoparticles must be separated from the surrounding matrix and then analyzed, whereas in other cases they can be analyzed *in situ*. In general, a specific set of analytical methods is required for each type of nanomaterial analyzed. For this reason, only a few examples are given here to illustrate the types of approaches that may be used.

The total protein content of nanomaterials can be determined by measuring their nitrogen content using Kjeldahl or Dumas methods. Alternatively, if the proteins can be solubilized, then the protein content can be determined using various UV-visible spectrophotometry methods, including the direct, Biuret, Lowry, and Bradford methods. The types and concentrations of different kinds of proteins and polysaccharides in nanoparticles can often be determined using gel electrophoresis (like SDS-PAGE) or size-exclusion chromatography (gel permeation) methods. The lipid content of nanoparticles can often be determined using solvent extraction methods, such as the Soxhlet method. The fatty acid composition of the triacylglycerols or phospholipids used to construct lipid-based or phospholipid-based nanoparticles can be determined by gas chromatography. The types and concentrations of vitamins, nutraceuticals, or other active agents encapsulated in nanoparticles can be measured using gas chromatography (GC) or high-performance liquid chromatography (HPLC) methods. The mineral composition of organic and inorganic nanoparticles can be determined by converting them into an ash (e.g., using a muffle furnace) and then using atomic absorption or emission spectroscopy to determine the concentration of the different elements present. Alternatively X-ray methods can be used to provide information about mineral ion concentrations in nanomaterials. Insights into the spatial distribution of the mineral elements in a nanomaterial can be obtained using sophisticated analytical instruments that combine electron microscopy with X-ray methods. Mass spectrometry, often combined with chromatography, can be used to provide information about the composition of certain kinds of nanoparticles. Information about the moisture, protein, carbohydrate, and lipid content can sometimes be obtained rapidly and simultaneously using specialized Fourier Transform infrared spectroscopy instruments; however, the instrument must be calibrated using numerous samples of known composition prior to application.

2.3 Particle concentration

2.3.1 Importance

The concentration of nanomaterials incorporated into a food or packaging material can be varied to obtain different effects. In general, adding nanomaterials will alter the physicochemical and functional properties of the materials they are incorporated into, including their optical, rheological, stability, barrier, flavor, and release/retention characteristics. For instance, the turbidity and viscosity of a liquid tend to increase as the concentration of nanoparticles added increases. Food manufacturers can usually control the concentration of nanoparticles they incorporate into their products by controlling the ingredients and processing operations used. However, it is often important to measure the concentration of nanoparticles present in foods, as this influences their physicochemical, sensory, and functional properties, as well as their potential toxicity. Measurements of nanoparticle concentration may also be required to ensure that a food or beverage product meets the appropriate government regulations. There is therefore a need to have suitable analytical techniques to determine nanoparticle concentrations in foods.

2.3.2 Characterization

A variety of analytical test methods are available for determining the concentration of nanoparticles in foods (McClements and McClements 2016). In some cases, it is necessary to separate the nanoparticles from the surrounding food matrix and then measure the amount present, which is often highly challenging. A variety of procedures may be required to isolate the nanoparticles from the food matrix, which depends on the nature of both the nanoparticles and the food matrix. Some of the procedures that may be used include physical, chemical, or enzymatic degradation of the food matrix (solid foods), as well as separation of the nanoparticles from the food matrix, e.g., by using gravitational forces, centrifugation, or filtration (liquid foods and digested solid foods). The nanoparticles can then be collected, washed, dried, and weighed. These approaches are most suitable for inorganic nanoparticles that may be resistant to chemical or enzymatic degradation.

In other cases, information about the nanoparticle concentration can be determined without isolating them. For instance, in nanoparticle suspensions the particle concentration can sometimes be measured using various scattering and microscopy approaches (see later). As an example, the concentration of nanoscale fat droplets in a nanoemulsion can be determined using light scattering methods. A laser beam is directed through a dilute nanoemulsion and the light intensity *versus* diffraction angle is measured. An appropriate mathematical model is then fitted to this light scattering pattern, which can be used to estimate the particle concentration and size.

2.4 Nanomaterial shape, aggregation state, and organization

2.4.1 Importance

The morphology, aggregation state, and location of the nanoparticles within a food or beverage product determine their impact on the physicochemical, sensory, and functional properties (McClements 2015b). For instance, the viscosity of a nanoparticle suspension tends to increase as the particle shape becomes more elongated or when the particles flocculate with each other. This is because the effective volume fraction of the disperse phase in the suspension increases, which leads to more energy dissipation during flow, thereby increasing the viscosity. Thus, a suspension of nanofibers tends to have a much higher viscosity than a suspension of nanospheres with the same concentration. Similarly, the viscosity of a nanoparticle suspension tends to increase when the nanoparticles associate with each other, especially when they link up to form a 3D network of nanoparticles that extends throughout the entire system. These phenomena can be used to create nanoparticle-based ingredients to modify the textural attributes of food and beverage products. The resistance of nanoparticles to gravitational separation, such as creaming and sedimentation, also depends on their aggregation state. For instance, a suspension of non-aggregated nanoparticles may be highly resistant to gravitational separation because the individual particles are so small that the gravitational forces are very weak. However, once the nanoparticles aggregate with each other they form much bigger particles that tend to cream or sediment at a much higher velocity, thereby leading to rapid phase separation. It should be noted, however, that if the nanoparticle concentration is sufficiently high, then particle aggregation may lead to the formation of a particle network that inhibits gravitational separation by preventing their movement. The aggregation state of digestible nanoparticles within the human gastrointestinal tract may also impact their rate and extent of digestion, as well as the bioavailability of any encapsulated hydrophobic bioactive substances (Tan and McClements 2021). For instance, lipid droplets that are aggregated when entering the small intestine are digested at a slower rate than non-aggregated ones because a lower lipid surface area is exposed to the lipases in the gastrointestinal fluids, thereby inhibiting lipid digestion and bioactive release and solubilization.

The location of the nanoparticles in a food may also be important. For instance, in an emulsion, any nanoparticles present may be dispersed throughout the aqueous phase, or they may be absorbed to the oil droplet surfaces, which influences the stability of the droplets to aggregation and gravitational separation. Emulsions containing particle-coated oil droplets, which are sometimes referred to as Pickering emulsions, often have a much greater resistance to coalescence than emulsions containing emulsifier-coated oil droplets (Berton-Carabin and Schroen 2015). Knowledge of the location of nanoparticles is also important when assessing their gastrointestinal fate and potential toxicity. For instance, the internalization of ingested nanoparticles by the epithelium and M-cells lining the gastrointestinal tract may determine their toxicity and behavior within the human body (McClements et al. 2017). The extent of internalization by these cells is strongly influenced by the size, shape, and aggregation state of the ingested nanoparticles within the gastrointestinal tract. Consequently, it is important to have appropriate analytical methods that can be used to quantify the shape, aggregation, and location of the nanoparticles in foods.

2.4.2 Characterization

Nanoparticles are too small to be observed directly because the unaided human eye only has a resolution of around 100 μm, which is about 100-fold larger than the biggest nanoparticles. Consequently, it is necessary to use microscopy instruments to provide information about the morphology, aggregation state, and location of nanoparticles. The most common microscopy methods used for this purpose are optical, electron, and atomic force microscopy (AFM), which are differentiated based on the physical principles used to obtain images of the samples being observed (Dudkiewicz et al. 2011, Bandyopadhyay et al. 2013, McClements and McClements 2016, Pan and Zhong 2016, Falsafi et al. 2020, Nizamov et al. 2022). Optical microscopy is based on the interactions of light waves with nanomaterials, whereas electron microscopy is based on the interactions of an electron beam with nanomaterials. In contrast, AFM is based on measuring the forces acting on an extremely sensitive probe that is moved across the surface of the sample. Each microscopy method can be characterized by three key attributes, its resolution, magnification, and contrast. *Resolution* is related to the smallest object that can be successfully observed using the microscope. *Contrast* is the ability of the microscope to distinguish the objects in an image from the background and from other objects. *Magnification* is the number of times the image of an object is larger than the actual object being observed. Modern microscopes are usually linked to computers that capture, store, and process digital versions of the image. Specialized image processing software can then be used to provide information about the size, shape, aggregation state, location, and sometimes composition of nanoparticles. In the remainder of this section, a brief overview of the main microscopy methods used to obtain this kind of information is given.

2.4.2.1 Optical microscopy

Optical microscopy instruments utilize the interactions between light waves and materials to obtain images (Falsafi et al. 2020, Nizamov et al. 2022). A typical modern optical microscope consists of a light source, a series of lenses, and a digital camera. The lenses direct the light waves through the sample and magnify the image obtained, which is then detected, displayed, and analyzed. An optical microscope typically has a *resolution* of around 500 nm, which is governed by the wavelength of light, the physical design of the device, and the Brownian motion of small particles. As a result, optical microscopes are unsuitable for characterizing the size or shape of most nanoparticles because they have dimensions below this limit. However, they can provide information about the degree of aggregation in nanoparticle suspensions since the aggregates formed often have dimensions above the limit of resolution. The *contrast* between nanoparticles and their surroundings in conventional optical microscopy is mainly due to differences in their refractive indices. The contrast can be enhanced by using phase contrast, polarization, or staining methods. Polarization is typically used for nanoparticles that have crystalline structures, whereas staining is used to distinguish the location of different components in a suspension (e.g., proteins, lipids, or carbohydrates). Laser scanning confocal microscopy (LSCM) is a powerful method for obtaining high-quality two- and three-dimensional images of nanoparticle suspensions. Several other kinds of advanced microscopy methods have also been developed that can provide information about the chemical composition of the different components in an image, such as infrared or Raman microscopy.

2.4.2.2 Electron microscopy

Electron microscopy is commonly utilized to determine the size, shape, and aggregation state of nanoparticles because it has a much better resolution than optical microscopy (Falsafi et al. 2020). Electron microscope instruments use electron beams, rather than light beams, to obtain images of nanomaterials. Nevertheless, the design of these instruments is somewhat like that of optical microscopes. In electron microscopy, an electron beam is directed through a series of magnetic fields that direct the beam at the sample and magnify the resulting image (Figure 2.2). The resolution of electron microscopy is much higher than that of light microscopy because the wavelength of electron beams is much smaller than that of light beams. In some powerful electron microscopy instruments, the *resolution* may be as low as a nanometer or less. The *contrast* in the images acquired using an electron microscope is mainly a result of differences in the electron densities of different components in the sample, which can be increased by using electron-dense heavy metal stains that bind to specific components. The *magnification* of an electron microscope is governed by the design of the instrument used. In the most powerful instruments, an image may be several million times larger than the object observed.

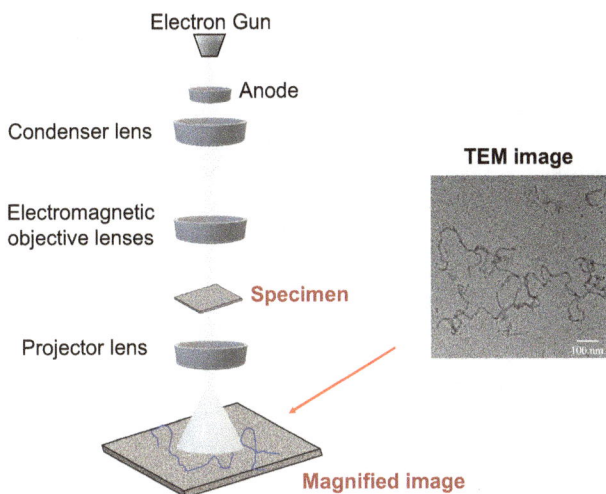

Figure 2.2: Schematic representation of a transmission electron microscope (TEM) that can be used to provide images of nanomaterials. The design of the instrument is like a light microscope, but an electron beam rather than a light beam is used. The TEM image is of protein nanofibers formed by controlled heating of β-lactoglobulin (kindly supplied by Charmaine Koo).

There are two main kinds of electron microscopy instrument that can be used to characterize the properties of nanomaterials: transmission and scanning electron microscopy (TEM and SEM) (Murphy 2012, Egerton 2016, Falsafi et al. 2020). TEM provides a 2D image by measuring the intensity of electron beams that have passed through different regions of the sample. SEM provides a more 3D image by measuring the intensity of electron beams scattered from the surfaces of a sample. Typically, TEM has a higher resolution and magnification than SEM, but the former requires more extensive sample preparation. In general, one must be careful when interpreting electron microscopy images of nanomaterials because sample preparation can cause appreciable alterations in their structural properties. For instance, samples may undergo fixation, dehydration, slicing, sputter-coating, and/or freezing before analysis, which can severely alter their structures.

2.4.2.3 Atomic force microscopy

Atomic force microscopy (AFM) instruments are also widely used to provide information about the structure of nanomaterials (Morris et al. 1999, Sitterberg et al. 2010, Ho et al. 2022). These instruments generate images of nanomaterials by scanning a small probe across their surfaces and measuring the magnitude of the interactions at different positions in the x-y direction (Figure 2.3). At small separations between the probe and the sample surface the interaction strength is relatively strong, whereas at large separations it is relatively weak. When the interaction is

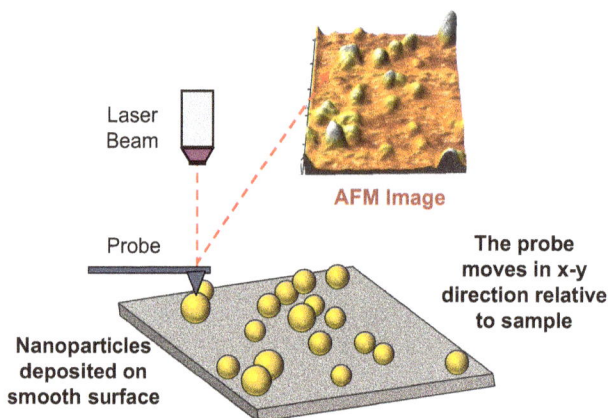

Figure 2.3: Schematic representation of an atomic force microscope (AFM) capable of providing information about the size, morphology, and aggregation state of nanoparticles. The AFM image is of protein-polysaccharide nanoparticles (kindly provided by Owen Jones).

strong, the probe is deflected to a greater distance. Typically, a laser is used to measure the deflection of the probe, which therefore provides a measure of the surface topology of the nanomaterial being tested. Alternatively, the force that the instrument must apply to keep the distance between the probe and the surface of the material constant is measured and reported. The *resolution* of an AFM instrument depends on the dimensions of the probe and how accurately it can be positioned relative to the surface of the sample. It is common for lateral resolutions (x-y direction) of around a few nanometers and vertical resolutions (y-direction) of around tens of nanometers to be obtained. An advantage of AFM for some applications is that it can provide information about the microstructure of both wet and dried samples.

2.4.3 Practical considerations

With all microscopy methods, it is important to carefully prepare the samples prior to analysis to avoid introducing any artefacts into the images. Typically, optical microscopy requires the least sample preparation, but it can only observe relatively large objects and so is unsuitable for observing individual nanoparticles. For most kinds of nanoparticles, electron microscopy is the most suitable method because it has sufficient resolution and magnification to provide clear images of the samples being analyzed. However, it may be necessary to selectively stain the samples to obtain sufficient contrast (e.g., lipids are usually stained with osmium tetroxide to distinguish them from other components, while proteins and lipids can be stained using uranyl acetate). The main challenge using electron microscopy is that the instruments are often relatively expensive and sample preparation can alter the structures being observed.

2.5 Particle size

2.5.1 Importance

The dimensions of nanomaterials, especially nanoparticles, are important because they affect their functional properties and their impact on food properties, including their appearance, texture, shelf-life, retention/release, and gastrointestinal fate (McClements 2015b). Some of the most important ways that the size of nanoparticles influences their physicochemical and functional properties are listed below:

Appearance: The impact of nanoparticles on the optical properties of foods, beverages, and packaging materials is strongly linked to their size. The turbidity or opacity of a suspension of particles tends to increase with increasing particle size until the wavelength of light is reached and then it decreases (Figure 2.4). The wavelength of light ranges from around 380 to 780 nm depending on its color. Consequently, the maximum scattering from a particle suspension occurs when the particle diameter is around a few hundred nanometers. In applications where a clear food or beverage product is required, it is important to use particles with diameters that are about tenfold smaller than the wavelength of light (< 50 nm) because then the light scattering is relatively weak. Conversely, in applications where

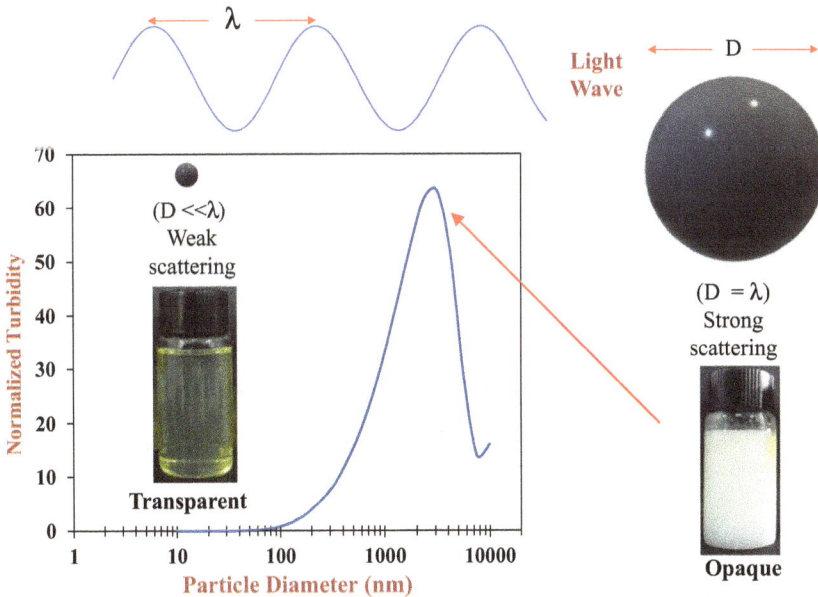

Figure 2.4: The impact of nanoparticles on the optical properties of foods and beverages depends on their dimensions relative to the radius of light. When nanoparticles are much smaller than the wavelength of light (d < 50 nm) they scatter light weakly and appear clear.

a cloudy or opaque product is required, it is better to use particles with diameters around a few hundred nanometers, so the light scattering is relatively strong.

Texture: In some products, the size of the nanoparticles has a pronounced impact on their rheological properties (McClements 2015b). This is usually the case when the nanoparticle concentration is relatively high, so that particle-particle interactions become important. Typically, the smaller the particle size, the closer the particles are together, and the stronger the particle-particle interactions. As a result, reducing the size of the particles in a suspension can cause an appreciable increase in viscosity and/or gel formation, even when the particle concentration is kept constant. Particle size is especially important in systems where there is a strong attractive or repulsive interaction between the particles (Figure 2.5). The impact of particle size on rheology may be utilized to modify the textural attributes of foods.

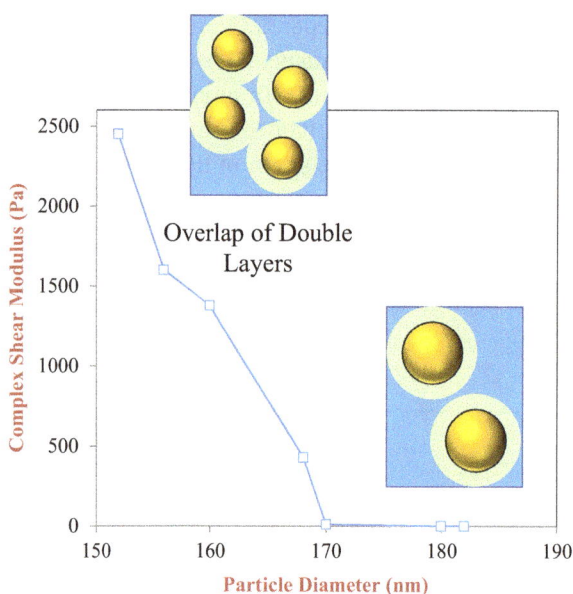

Figure 2.5: Impact of particle size on the shear modulus of SDS-stabilized oil-in-water nanoemulsions. The nanoemulsions have some elastic-like properties when the particle size falls below a critical level due to overlap of the electrical double layers.

Shelf-life: The size of nanoparticles also has a major impact on the shelf-life of the food and beverage products they are incorporated into. Typically, the resistance of nanoparticles to gravitational separation (creaming or sedimentation) or aggregation (flocculation, coalescence, or coagulation) increases as their dimensions decreases. Consequently, it may be important to ensure that the size of the individual nanoparticles is relatively small and that they do not tend to aggregate (which increase the particle size).

Retention/release: The retention and release of encapsulated substances contained within nanoparticles depend on their size. Typically, the smaller the particle size, the lower the retention, and the faster the release. This is important when designing nanoparticle-based delivery systems for the controlled release of bioactive agents, such as antimicrobials, nutraceuticals, flavors, or vitamins.

Gastrointestinal fate: The behavior of nanoparticles within the gastrointestinal tract also depends on their size (McClements et al. 2017, McClements 2020). Small non-digestible nanoparticles may be able to travel through the gastrointestinal tract, penetrate the mucus layer that coats the intestinal lining, and be absorbed by the epithelium cells (Figure 2.6). In contrast, large individual particles or nanoparticle aggregates may not be able to penetrate the mucus layer, which has a pore size around 400 nm. The fate of digestible nanoparticles in the human gut also depends on their dimensions. Typically, smaller nanoparticles are digested more rapidly than larger ones because they have a larger specific surface area for the digestive enzymes to adsorb to. As a result of their faster and more complete digestion, smaller nanoparticles often give a higher bioavailability of encapsulated substances than larger ones. In the case of lipid nanoparticles, such as the small fat droplets in nanoemulsions, the bioaccessibility of hydrophobic nutraceuticals and vitamins increases as the particle size decreases because more of them are released from the digested oil droplets and solubilized in the mixed micelles formed.

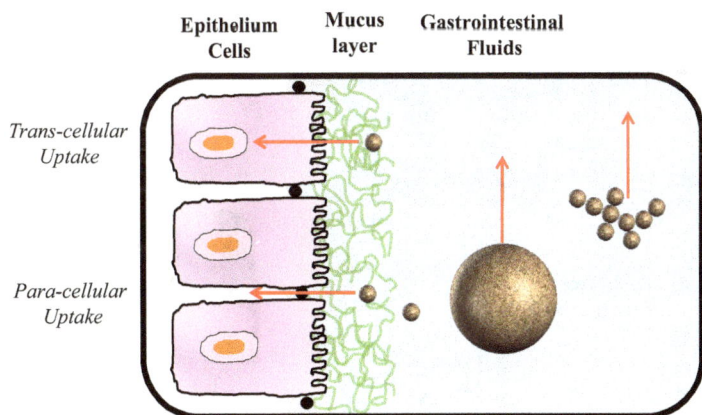

Figure 2.6: Small non-digestible nanoparticles may be able to travel through the mucus layer lining the intestinal walls and be absorbed by epithelium cells, whereas large ones or aggregated ones cannot.

2.5.2 Characterization

Given the importance of nanoparticle dimensions on their functional properties it is important to be able to measure their particle size characteristics. Typically, the full particle size distribution (PSD), mean particle diameter, and polydispersity index are measured (McClements and McClements 2016). Several kinds of particle sizing instruments are commercially available for this purpose, which are based on different physical principles (McClements 2007). In some applications, the size of the particles can be determined using microscopy instruments, such as the optical, electron, and atomic force microscopy methods discussed earlier. However, this is often a time-consuming and laborious process since it is necessary to prepare many samples and take many images to get statistically reliable data. More commonly, analytical instruments specifically designed to measure the particle size of colloidal dispersions, like nanoparticles, are used. These instruments are relatively simple to use, rapid, and give reliable and reproducible data (provided the measurements are carried out carefully). The most widely used particle sizing instruments for analyzing nanoparticles are based on light scattering: dynamic and static light scattering. Static light scattering, which is also known as laser diffraction, is based on transmitting a laser beam through a dilute nanoparticle suspension and measuring the light scattering pattern, i.e., the light intensity *versus* diffraction angle (Figure 2.7). A mathematical model is then used to convert the scattering pattern into a particle size distribution. This kind

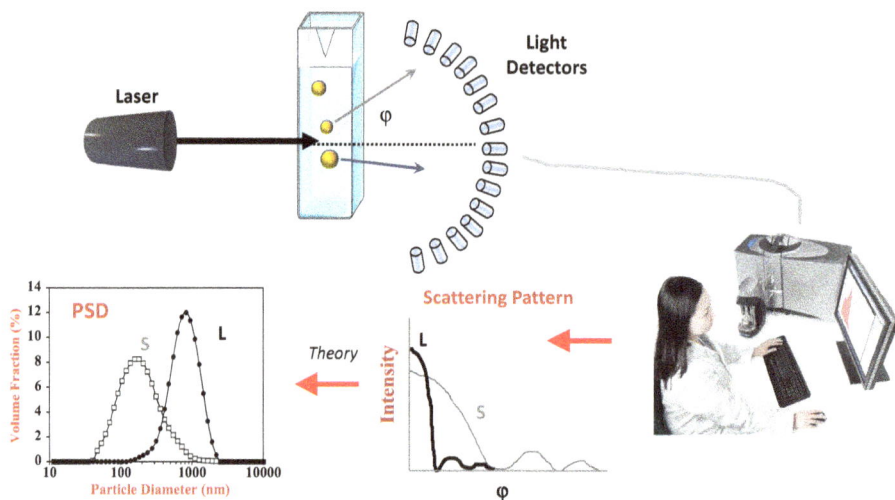

Figure 2.7: The particle size distribution (PSD) of nanoparticle suspensions can be measured using static light scattering (laser diffraction), which measures the scattering pattern (intensity *versus* diffraction angle) and then uses a mathematical model to convert the data into a particle size distribution. (Image of cuvette from Servier, Medical Arts, Creative Commons 3.0; Image of instrument kindly provided by Malvern Panalytical).

of instrument is suitable for providing information about the size of particles that have diameters ranging from around 100 nm to 1000 μm. It is therefore only suitable for analyzing relatively large individual nanoparticles or nanoparticle aggregates. Dynamic light scattering is based on measurements of the fluctuation in light intensity over time when a laser beam is transmitted through or reflected off a nanoparticle suspension (Figure 2.8). A mathematical model is then used to relate the frequency distribution of these fluctuations into a particle size distribution. This kind of instrument provides information about the size of particles with diameters ranging from about 1 nm to 5 μm. As a result, it is more suitable than static light scattering at characterizing the dimensions of small nanoparticles. Conversely, it is unsuitable for analyzing clusters of nanoparticles with diameters above this range.

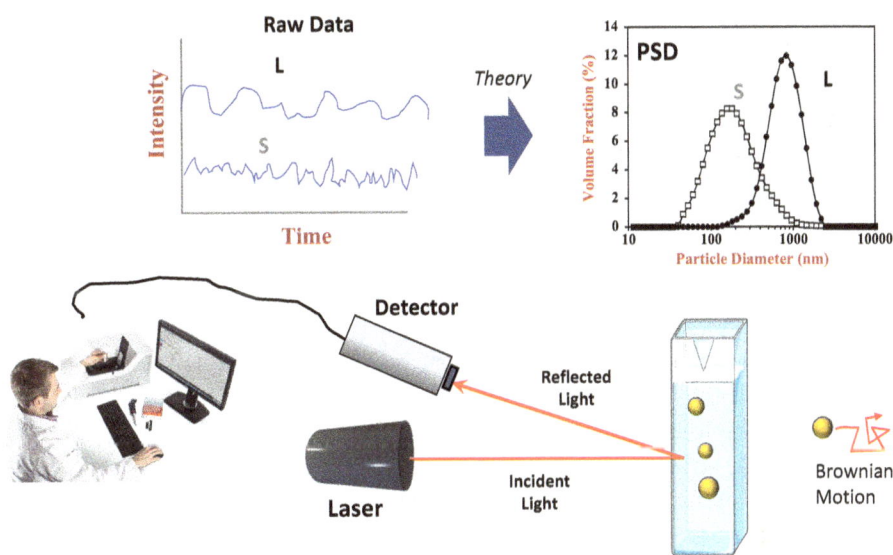

Figure 2.8: The particle size distribution (PSD) of nanoparticle suspensions can be measured using dynamic light scattering, which measures fluctuations in the scattering pattern of light reflected from the surface of a nanoparticle suspension and then uses a mathematical model to convert the data into a particle size distribution. (Image of cuvette from Servier, Medical Arts, Creative Commons 3.0; Image of instrument kindly provided by Malvern Panalytical).

Some knowledge of the properties of the sample being tested is required by the light scattering instrument to calculate the particle size distribution using the mathematical models. For instance, the refractive index and absorption coefficient of the nanoparticles and surrounding liquid are required for static light scattering, whereas the viscosity of the surrounding liquid is also required for dynamic light scattering. Several other kinds of analytical instruments are also available for measuring nanoparticle size characteristics, including centrifugation, electrical pulse counting, and neutron scattering methods, but these are much less commonly used (McClements and McClements 2016).

2.5.3 Practical considerations

Commercial particle sizing instruments are designed to be simple to use and can quickly generate information about the particle size distribution, mean particle diameter, and polydispersity index of a sample. Even so, sample preparation and analysis must be carefully carried out to obtain reliable information about the size of nanoparticles (McClements 2015a, McClements and McClements 2016). Some of the most important factors that must be considered are listed below:

- Samples should be diluted with a solution with the same pH (and ideally ionic strength) as the original sample; otherwise, there may be changes in the aggregation state of the nanoparticles.
- Stirring conditions during sample preparation and analysis should be designed so they do not impact the aggregation state of the nanoparticles.
- An appropriate mathematical model should be selected to interpret the experimental measurements. For aggregated samples, the data obtained should be treated with caution because the mathematical models used by most instruments assume the particles are homogeneous isolated spheres, which is not the case for clusters of nanoparticles.
- Accurate physicochemical properties (such as refractive index, absorption, and viscosity) of the nanoparticles and/or surrounding fluid should be used in the instrument software.
- The particle sizing instrument should be maintained and operated correctly. For instance, it is important to ensure that the measurement cells are optically clear, with no scratches or dirt that can scatter light and affect the results.

2.6 Particle charge

2.6.1 Importance

The functional performance and impact of nanoparticles on the properties of foods and beverages often depend on their electrical characteristics (McClements 2015b). Nanoparticles often have an electrical charge because they have charged groups present at their surfaces, e.g., ionized chemical groups that are part of the particle core or adsorbed surfactants, proteins, polysaccharides, or phospholipids that are ionized. Overall, the electrical properties of nanoparticles are determined by the sign, number, and location of the charged groups on their surfaces, as well as the composition of the surrounding solution (such as the number and type of ions present, as well as its dielectric constant). Nanoparticles may have charges that go from strongly positive to strongly negative depending on their surface chemistry and solution conditions. The sign and magnitude of the charge on a nanoparticle plays an important role in determining its stability and its interactions with its environment. A high electrical charge can often prevent nanoparticles aggregating with their neighbors by generating a strong electrostatic repulsion between them. However, charged nanoparticles may be attracted to oppositely charged substances in their environment, such as mineral ions, biopolymers, or particles, which can promote their aggregation through charge neutralization and bridging mechanisms. Moreover, the adsorption of cationic transition metals to the surfaces of anionic nanoparticles can promote oxidation of the substances inside them (such as polyunsaturated lipids). The electrical characteristics of nanoparticles may also influence their tendency to stick to surfaces they are exposed to, such as processing equipment (e.g., mixers or pipes) or containers (e.g., cans, bottles, or caps). The gastrointestinal fate of nanoparticles may also depend on their electrical characteristics by altering their interactions with the mucus layer that coats the human gut or with digestive substances like mucin, enzymes, and bile salts. Consequently, there is a need for instrumental methods to characterize the electrical characteristics of nanoparticles (McClements and McClements 2016).

2.6.2 Characterization

The electrical characteristics of nanoparticles are usually quantified by measuring the effective surface potential (ζ-potential) as a function of solution pH (Hunter 1986, McClements 2015a). The ζ-potential depends on the net number of charged groups per unit surface area (surface charge density) of the nanoparticles, as well as the ionic composition (mineral ion valence and concentration) of the surrounding liquid. Typically, the ζ-potential increases as the net surface charge increases and as the ionic strength of the surrounding liquid decreases. The most common instruments used to measure the ζ-potential of nanoparticles are based on particle electrophoresis (McClements and McClements 2016).

Particle electrophoresis instruments typically use a laser beam to determine the speed and direction that charged nanoparticles move when an electrical field is applied across the sample using two electrodes (Figure 2.9). The nanoparticles are typically dispersed within an aqueous solution of known pH and ionic composition prior to analysis. After the electric field is applied, the charged nanoparticles are attracted toward the oppositely charged electrode at a velocity that depends on their surface charge: the higher the surface charge, the faster they move. The velocity also depends on the viscosity of the surrounding liquid (the higher the viscosity, the slower they move), and so this should be known and controlled during the experiment. The software in the instrument uses the measurements of the nanoparticle velocity and direction of movement to calculate their electrophoretic mobility, which is then converted into a ζ-potential. Nanoparticle suspensions usually must be diluted before analysis to avoid particle-particle interaction effects.

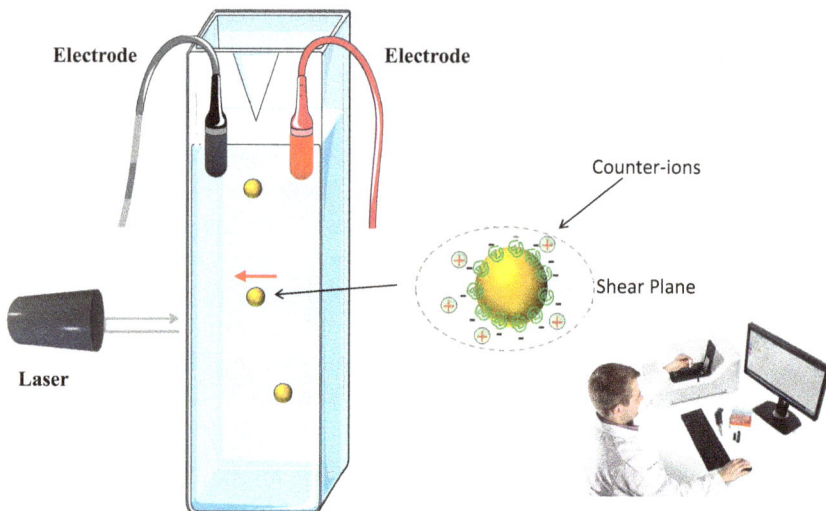

Figure 2.9: The effective surface potential of nanoparticles can be measured using microelectrophoresis instruments. The direction and velocity that the nanoparticles move are determined using a suitable technique (such as a laser) when a well-defined electrical field is applied across the measurement cell. Highly schematic and oversimplified representation of microelectrophoresis device. (Image of instrument kindly provided by Malvern Panalytical).

2.6.3 Practical considerations

It is critical that sample preparation and instrument operation is carried out correctly to obtain reliable ζ-potential measurements of nanoparticles (McClements 2007, McClements 2015a). Many of the same factors that affect light scattering measurements (see earlier) also affect ζ-potential measurements. For instance, it is important to dilute samples with a solution that has the same pH, and ideally ionic strength, as the original sample. Moreover, it is important to input the correct physicochemical parameters of the nanoparticles and surrounding liquid into the instrument software, such as the refractive indices of the nanoparticles and liquid, as well as the shear viscosity of the liquid. Finally, it is important that the mathematical model used by the instrument is correct. This often depends on the size of the particles relative to the range of the electric field around them (which is described by the Debye screening length).

2.7 Particle physical state

2.7.1 Importance

The nanoparticles in foods may be solid, liquid, or semi-solid, which influences their functional performance and impact on food properties (McClements 2015b). Moreover, the physical state of a nanoparticle may change when environmental conditions are altered, e.g., they may melt or crystallize when the temperature is raised or lowered, or they may undergo polymorphic transitions. Food-grade nano-particles containing biopolymers (like proteins and polysaccharides) may undergo conformational changes at specific temperatures (such as thermal denaturation or helix-coil transitions). Some types of polymeric nanoparticles may also undergo glass-rubbery transitions when they are heated, or when the water activity is increased. These kinds of transitions can greatly alter the physical stability of nano-particles, as well as their ability to retain and release encapsulated components. For instance, lipid nanoparticles may aggregate due to partial coalescence when they are partially crystalline because a crystal in one particle sticks into a liquid region in another particle. The ability of nanoparticles to undergo solid-to-liquid phase changes upon heating can be used as a triggered release mechanism. An en-capsulated substance remains inside the nanoparticles at low temperatures where they are solid but is then released when they are heated above their melting point. For these reasons, it is important to have analytical instruments to characterize the physical state of nanoparticles (McClements and McClements 2016). Several different kinds of analytical methods are available to obtain this kind of information, with the most suitable one depending on the nature of the nanoparticles and phase transitions being studied.

2.7.2 Characterization

2.7.2.1 Dilatometry

Information about the physical state of some kinds of nanoparticles can be deter-mined by measuring changes in their density, which is referred to as dilatometry (McClements 2015b). Typically, the density of a nanoparticle suspension is mea-sured as a function of temperature or time. The density of a solid phase is usually higher than that of a liquid phase, which means that melting and crystallization of nanoparticles can be measured using this method. For instance, the density of a suspension of lipid nanoparticles increases when it is cooled below the crystalliza-tion temperature of the lipid phase but decreases when it is heated above the melt-ing temperature (Figure 2.10).

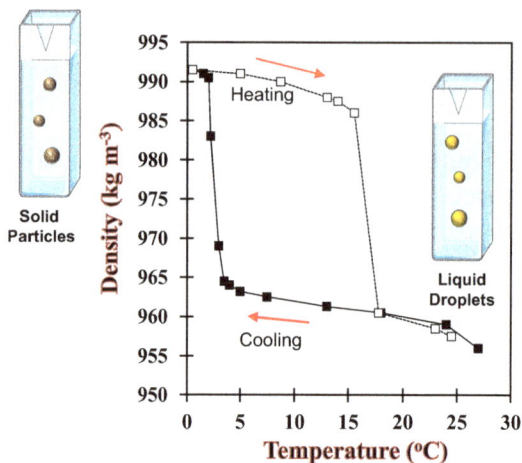

Figure 2.10: Measurements of the density of nanoparticle suspensions (hexadecane oil-in-water nanoemulsions) versus temperature can be used to monitor crystallization and melting behavior.

2.7.2.2 Differential scanning calorimetry

Differential scanning calorimetry (DSC) is one of the most common methods of monitoring phase transitions of nanoparticles (Qiao et al. 2017). This method is based on measuring the heat released or absorbed by the nanoparticles when they are heated or cooled at a controlled rate. The nanoparticles may be in a powdered form, or they may be suspended within an appropriate fluid. Experimentally, the nanoparticles are placed within a sample pan and an empty pan is used as a reference. The heat flow required to keep the two pans at the same temperature when they are heated or cooled at a known rate is then measured, and then an enthalpy change *versus* temperature profile is calculated. DSC is suitable for characterizing different kinds of nanoparticle phase transitions, such as crystallization, melting, thermal denaturation, glass-rubbery transitions, and polymorphic transformations. Typically, the transition temperatures are determined from the position of the peaks, while the transition enthalpies are determined from the area under the peaks (Figure 2.11).

2.7.2.3 Nuclear magnetic resonance

Nuclear magnetic resonance (NMR) instruments can also be used to provide information about changes in the physical state of nanoparticles (McClements 2015a). The sample to be analyzed is placed within a powerful static magnet, which causes the hydrogen nuclei to be aligned with the magnetic field. A radiofrequency pulse is then applied to the sample that generates a transitory magnetic field that is perpendicular to the original static one. This pulse causes the hydrogen nuclei to become re-aligned in the perpendicular direction. The instrument then measures the

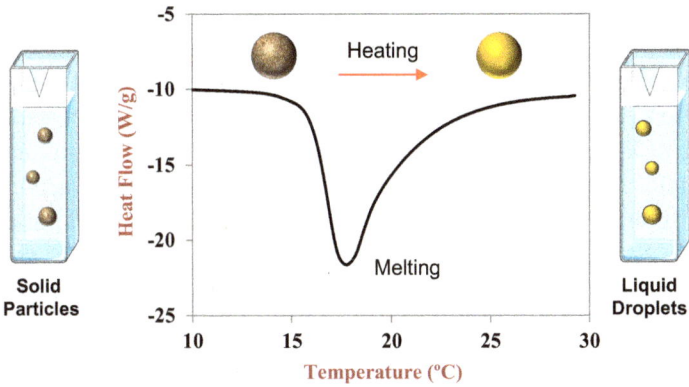

Figure 2.11: Measurements of the enthalpy change of nanoparticle suspensions (hexadecane oil-in-water nanoemulsions) versus temperature using differential scanning calorimetry can be used to monitor crystallization or melting behavior.

rate of decay of the hydrogen nuclei back to their original alignment. The decay rate is governed by the physical state of the material being analyzed, being much faster for solids than liquids (Figure 2.12). As a result, the physical state of the nanoparticles within a test sample can be determined by measuring the rate of decay of the NMR signal after the radiofrequency pulse is removed.

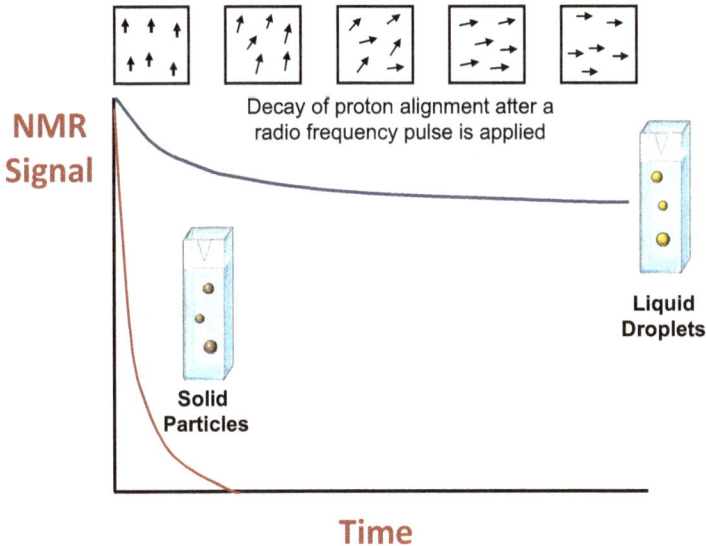

Figure 2.12: Measurements of the decay of the NMR signal with time can be used to monitor phase transitions. The decay rate is much faster for solids that for liquids.

2.7.2.4 X-ray diffraction analysis

Information about the physical state of nanoparticles can be obtained using X-ray diffraction methods (Hartel 2013, Qiao et al. 2017). Typically, the material to be analyzed is converted into a powdered form and then held at the required measurement temperature. A beam of X-rays is then passed through the sample and the resulting diffraction pattern (intensity *versus* angle) is measured. Analysis of the diffraction pattern gives useful information about the organization of the molecules within a material. For instance, they can be used to ascertain whether the nanoparticles are in an amorphous or crystalline state, and if they are in a crystalline state, they can provide information about the polymorphic form.

2.7.2.5 Microscopy

Some microscopy instruments can provide insights into the physical state of nanoparticles (Falsafi et al. 2020). For example, optical microscopes fitted with crossed polarizers can provide information about the presence of ordered regions within a sample, such as crystals or liquid crystals. In some cases, transmission electron microscopy (TEM) can give insights into the packing of the molecules within nanoparticles, such as triacylglycerol molecules in solid lipid nanoparticles (Bunjes et al. 2007, Kuntsche et al. 2011).

2.7.3 Practical considerations

Several factors must be considered when identifying a suitable instrument to provide information about the physical state of nanoparticles. The instrument must be capable of detecting the type of phase transition occurring in the nanoparticles, such as melting/crystallization, polymorphic, glass-rubbery, thermal denaturation, or sol-gel transitions. The instrument must be able to cover the range of temperatures where the transition occurs. The sample preparation methods used should not change the physical state of the nanoparticles prior to analysis. In complex systems, the instrument must distinguish the phase transitions of the nanoparticles from those of any other constituents that might be present.

2.8 Structure-function relations

Food and agricultural companies would like to utilize nanomaterials to create innovative products or to improve the performance of existing products. Ultimately, the functional performance of nanomaterials depends on their structural and physicochemical properties, like composition, size, charge, shape, physical state, and aggregation state. Consequently, it is important for companies to define what functional attributes they require and then design nanomaterials with characteristics that can provide these attributes. Numerous examples of nanoparticle design for specific applications will be provided in the following chapters. For this reason, only a few brief examples are given here. Nanoparticles can be designed to encapsulate hydrophobic substances so that they can be incorporated into clear foods and beverages. This is achieved by using nanoparticles with a hydrophobic interior and hydrophilic exterior. Moreover, the nanoparticles must be sufficiently small (< 50 nm) to only scatter light weakly, so they do not cause the product to look cloudy or opaque. Nanoparticles can also be designed to increase the bioavailability of hydrophobic substances (such as vitamins and nutraceuticals). In this case, it is important to ensure they are rapidly digested in the human gut and release the hydrophobic substances. In addition, they may contribute to the solubilization and transport of these substances to the epithelium cells where they can be absorbed. In general, it is important to develop structure-function relationships that link the properties of nanomaterials to their functional performance. These relationships can then be used to guide the formulation of innovative nanomaterials with the properties required for specific applications.

2.9 Summary

In this chapter, the most important characteristics of food-grade nanoparticles have been reviewed, as well as methods for measuring them. Knowledge of nanoparticle properties is important for designing advanced materials with the required functional attributes, such as appearances, mechanical properties, retention/release behavior, stability, and gastrointestinal fate. Research is still required to develop robust structure-function relationships that relate nanoparticle properties to the functional performance of these advanced materials. Moreover, appropriate analytical instruments are required to characterize the properties of nanoparticles and other nanomaterials used by the food industry. These instruments are needed for research and development purposes when designing new materials, as well as for quality assurance purposes when manufacturing these materials at commercial scale to ensure they meet the desired specifications.

References

Bandyopadhyay, S., J. R. Peralta-Videa and J. L. Gardea-Torresdey (2013). "Advanced analytical techniques for the measurement of nanomaterials in food and agricultural samples: A review". Environmental Engineering Science 30(3): 118–125.
Berton-Carabin, C. C. and K. Schroen (2015). "Pickering emulsions for food applications: background, trends, and challenges" Annual Review of Food Science and Technology Vol 6. M. P. Doyle and T. R. Klaenhammer 6: 263–297.

Bunjes, H., F. Steiniger and W. Richter (2007). "Visualizing the structure of triglyceride nanoparticles in different crystal modifications.". Langmuir **23**(7): 4005–4011.

Dudkiewicz, A., K. Tiede, K. Loeschner, L. H. S. Jensen, E. Jensen, R. Wierzbicki, A. B. A. Boxall and K. Molhave (2011). "Characterization of nanomaterials in food by electron microscopy". Trac-Trends in Analytical Chemistry **30**(1): 28–43.

Egerton, R. (2016). Physical Principles of Electron Microscopy: An Introduction to TEM, SEM, and AEM. New York, N.Y., Springer Science.

Falsafi, S. R., H. Rostamabadi, E. Assadpour and S. M. Jafari (2020). "Morphology and microstructural analysis of bioactive-loaded micro/nanocarriers via microscopy techniques; CLSM/SEM/TEM/AFM". Advances in Colloid and Interface Science **280**: 1–15.

Hartel, R. W. (2013). "Advances in food crystallization". Annual Review of Food Science and Technology Vol 4. M. P. Doyle and T. R. Klaenhammer. Palo Alto Annual Reviews **4**: 277–292.

Ho, T. M., F. Abik and K. S. Mikkonen (2022). "An overview of nanoemulsion characterization via atomic force microscopy." Critical Reviews in Food Science and Nutrition. **62**(18), 4908–4928.

Hunter, R. J. (1986). Foundations of Colloid Science. Oxford, Oxford University Press.

Kuntsche, J., J. C. Horst and H. Bunjes (2011). "Cryogenic transmission electron microscopy (cryo-TEM) for studying the morphology of colloidal drug delivery systems". International Journal of Pharmaceutics **417**(1–2): 120–137.

McClements, D. J. (2007). "Critical review of techniques and methodologies for characterization of emulsion stability". Critical Reviews in Food Science and Nutrition **47**(7): 611–649.

McClements, D. J. (2015a). Food Emulsions: Principles Practice, and Techniques. Boca Raton, CRC Press.

McClements, D. J. (2015b). Nanoparticle- and Microparticle-based Delivery Systems. Boca Raton, FL, CRC Press.

McClements, D. J. (2020). "Advances in nanoparticle and microparticle delivery systems for increasing the dispersibility, stability, and bioactivity of phytochemicals". Biotechnology Advances **38**: 1–13.

McClements, D. J., H. Xiao and P. Demokritou (2017). "Physicochemical and colloidal aspects of food matrix effects on gastrointestinal fate of ingested inorganic nanoparticles". Advances in Colloid and Interface Science **246**: 165–180.

McClements, J. and D. J. McClements (2016). "Standardization of nanoparticle characterization: methods for testing properties, stability, and functionality of edible nanoparticles". Critical Reviews in Food Science and Nutrition **56**(8): 1334–1362.

Morris, V. J., A. P. Gunning and A. R. Kirby (1999). Atomic Force Microscopy for Biologists. London, UK, Imperial College Press.

Murphy, D. B. (2012). Fundamentals of Light Microscopy and Electronic Imaging. New York, NY, John Wiley & Sons.

Nizamov, S., S. D. Sazdovska and V. M. Mirsky (2022). "A review of optical methods for ultrasensitive detection and characterization of nanoparticles in liquid media with a focus on the wide field surface plasmon microscopy". Analytica Chimica Acta **1204**: 1–32.

Pan, K. and Q. X. Zhong (2016). "Organic nanoparticles in foods: fabrication, characterization, and utilization." Annual Review of Food Science and Technology Vol 7. M. P. Doyle and T. R. Klaenhammer **7**: 245–266.

Qiao, Y. Q., R. R. Qiao, Y. N. He, C. P. Shi, Y. Liu, H. X. Hao, J. Su and J. Zhong (2017). "Instrumental analytical techniques for the characterization of crystals in pharmaceutics and foods." Crystal Growth & Design **17**(11): 6138–6148.

Sitterberg, J., A. Ozcetin, C. Ehrhardt and U. Bakowsky (2010). "Utilising atomic force microscopy for the characterisation of nanoscale drug delivery systems." European Journal of Pharmaceutics and Biopharmaceutics **74**(1): 2–13.

Tan, Y. B. and D. J. McClements (2021). "Improving the bioavailability of oil-soluble vitamins by optimizing food matrix effects: A review". Food Chemistry **348**: 1–12.

Chapter 3
Applications of nanotechnology in agriculture

3.1 Introduction

The global agricultural industry produces the edible substances we use to create our foods, such as fruits, vegetables, cereals, nuts, spices, herbs, meat, fish, egg, and milk. In the case of plant-based foods, the crops must be planted, cultivated, harvested, and stored prior to being processed into foods. In the case of animal-based foods, livestock animals must be raised and slaughtered, dairy cattle must be raised, housed, and milked, and laying hens must be raised and their eggs collected. The agricultural production system should generate enough food to feed the global population, which is close to 8 billion at present and predicted to grow to close to 10 billion by 2050 (ourworldindata.org). While achieving this goal, it is important that the agricultural system does not degrade our environment (such as land, water, and air) so that future generations can also meet their food production needs (Poore and Nemecek 2018). Moreover, the impact of the food production system on pollution, greenhouse gas emissions, and biodiversity loss should be minimized (Willett et al. 2019). There are several ways the current food production system could be optimized to achieve these goals, including raising agricultural productivity, reducing agricultural waste, making agricultural crops more resilient, and converting agricultural waste materials into value added products (WEF 2019).

Nanotechnology can be used to address several of the challenges facing the modern agricultural industry (He et al. 2019). In this chapter, several areas where nanotechnology could be employed to improve the productivity, increase the sustainability, reduce the waste, or decrease the pollution caused by the agricultural industry are highlighted.

3.2 Nanoenabled pesticides

A substantial amount of the crops and livestock raised for food are currently wasted because they are damaged or contaminated by various kinds of pests, including microbes (such as bacteria, molds, and viruses), insects (such as locusts, beetles, and worms), and animals (such as mice and rats) (Girotto et al. 2015, Kibler et al. 2018). The productivity of the agricultural system could therefore be enhanced by reducing the fraction of crops lost due to these pests. A wide variety of pesticides has been developed to treat different kinds of pests that may contaminate different types of agricultural crops. It has been reported that around 4 million tons of pesticides are used globally per year to treat agricultural crops (Usman et al. 2020). The judicious use of these pesticides improves yields and reduces waste. However, their

https://doi.org/10.1515/9783110788457-003

overuse can lead to environmental pollution, excessive occupational exposure, and food contamination, which can cause human health problems, including asthma, birth defects, cancer, diabetes, immune disorders, and obesity (Damalas and Eleftherohorinos 2011, Mostafalou and Abdollahi 2013, Yao et al. 2020). Moreover, the repeated use of pesticides can cause pests (especially microorganisms and insects) to develop resistance to them due to evolutionary pressures, thereby decreasing their efficacy. In addition, it has been reported that as much as 90% of applied pesticides are not having their intended effects, which is leading to environmental pollution and damage to human health (Velez et al. 2021). For these reasons, there is a focus on creating a new generation of pesticides that are efficacious but that cause fewer environmental and health concerns. In addition, there is growing interest in the design of smart pesticide formulations that can control the release of pesticides (Vejan et al. 2021). For instance, these formulations can be designed to give a prolonged pesticide release, or they may only release the pesticides when specific environmental conditions are experienced (such as a change in temperature, relative humidity, or pH). The utilization of these smart delivery systems may reduce the concentration of pesticides needed to treat crops, thereby reducing their undesirable environmental effects. In addition, nanopesticides may be designed to stick to the surfaces of agricultural crops more strongly than conventional formulations, thereby leading to fewer run-offs into the environment (Kah et al. 2019). Thus, nanotechnology has great potential for increasing the functional performance of conventional pesticides (Kah 2015, Grillo et al. 2016, McClements 2019, Singh et al. 2021).

The performance of nanoenabled pesticides can be controlled by manipulating the properties of the nanoparticles containing the pesticides, such as their composition, size, structure, aggregation state, and interfacial properties (Figure 3.1), which impacts their ability to penetrate into the interior of plants (Figure 3.2) (Kah 2015, Grillo et al. 2016, McClements 2019). Nanopesticides may be formulated from either inorganic or organic substances, such as copper oxides or essential oils.

3.2.1 Inorganic nanopesticides

Numerous studies have examined the potential of using inorganic nanopesticides to treat different kinds of pests on different kinds of crops. A few examples are given here:

–   *Copper nanoparticles*: Studies have shown that copper-based nanoparticles give better protection against fungal diseases on watermelons than conventional copper-based pesticides, with their effectiveness being influenced by nanoparticle properties like composition and dimensions (Elmer et al. 2018). Other studies have shown that copper nanoparticles are effective at combating fungal diseases that affect important agricultural crops without causing toxicity to the plants (Vanti et al. 2020). For instance, copper nanoparticles were able to

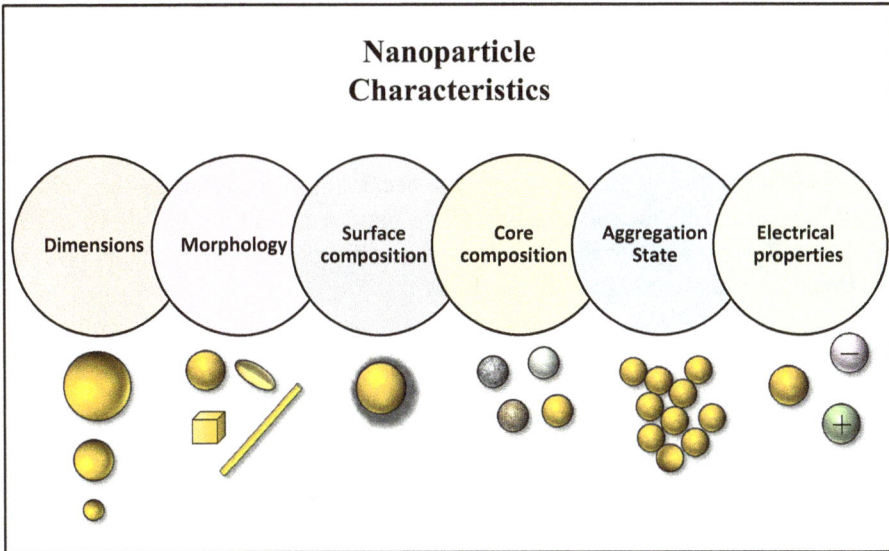

Figure 3.1: The effectiveness of nanopesticides and nanofertilizers depends on their physicochemical and structural characteristics.

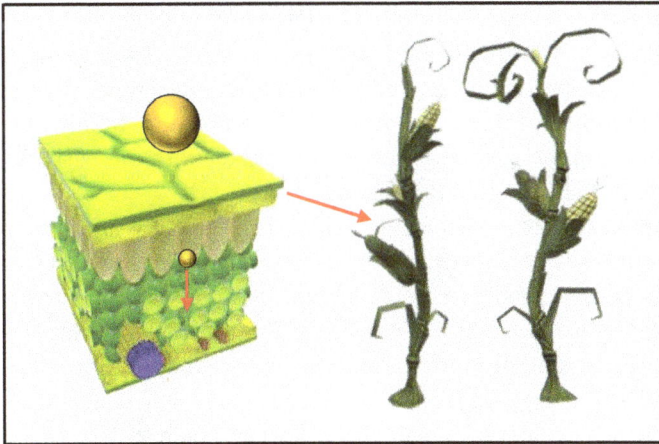

Figure 3.2: Nanoparticles internalized into plant cells may be used as sensors to provide information about the plant's maturity, health status, or water/nutrient needs.

inhibit the growth of *Rhizoctonia solani* and *Pythium aphanidermatum*, which cause a disease known as "damping off" in seeds and seedlings.

– *Silver nanoparticles:* Studies have also shown that silver-based nanoparticles can also be used to treat or prevent various kinds of plant pathogens. For instance, they have been used to inhibit the growth of *Xanthomonas axonopodis* and *Xanthomonas campestris*, which are common pathogens on some agricultural crops (Vanti et al. 2019). Silver nanoparticles have also been shown to exhibit strong antimicrobial activity against another common plant pathogen (*Ralstonia solanacearum*) that causes tobacco bacterial wilt (Chen et al. 2016). These authors showed that the nanoparticles disrupted the bacterial cell membranes and damaged key proteins in the bacteria, thereby reducing the viability of the pathogenic microorganisms. Silver nanoparticles have also been shown to be effective antimicrobial agents against fungal plant pathogens, such as *Monilia fructicola* (Malandrakis et al. 2020).

Several other kinds of inorganic nanoparticles have also been shown to exhibit pesticide activity against various kinds of pests that contaminate important agricultural crops, such as those based on silver, titanium, iron, silicon, zinc, magnesium, and aluminum (Wani et al. 2012, Shang et al. 2019). The ability of nanopesticides to effectively deactivate pests is often a result of the small size of the particles they contain. As the size of the particles decreases, their ability to penetrate into plants through the pores in their roots, leaves, and stems increases (Figure 3.2), which brings them closer to the location where pathogenic bacteria reside (Hong et al. 2021). The interfacial characteristics of nanopesticides can also be manipulated to improve their targeting to pests, as well as their ability to deactivate them (Kah et al. 2019). For instance, the electrical charge and hydrophobicity of the nanoparticle surfaces can be controlled so they are more likely to stick strongly to the surfaces of plants and pests. Cationic nanopesticides would be expected to strongly adhere to the anionic surfaces of plants or pests, thereby increasing their ability to bring the pesticides into close contact with the pests, as well as decreasing agricultural run-off of the pesticides.

When developing innovative nanoenabled pesticides it is important to consider their potential toxicity and negative impacts on the environment. There are several potentially adverse effects associated with using inorganic nanoparticles as pesticides (Chaud et al. 2021, Grillo et al. 2021, Xu et al. 2022). They may migrate into the environment, such as the surrounding land, water, or air, where they may accumulate or interact with various biological species. For instance, they may be taken up by plants, insects, or animals due to their small dimensions and high surface areas, and thus enter the human food supply chain (Chaud et al. 2021, Grillo et al. 2021, Xu et al. 2022). Studies have shown that treatment of agricultural crops may lead to changes in their nutritional profile, such as a reduction in essential amino acids or phenolic compounds (Zhao et al. 2016, Huang et al. 2018). The impact of these changes on human health and wellbeing are currently not well understood, but it will be important to optimize nanopesticide doses and types to effectively reduce contamination by pests without adversely altering the desirable nutrient content of the agricultural crops. It is also important to ensure that long-term use of nanopesticides does not cause damage to soil quality, such as the diversity of its microbiome

(Simonin et al. 2018, Peixoto et al. 2021). Clearly, further research is needed to establish the potential health and environmental consequences of using these kinds of nanopesticides on agricultural crops (Kah et al. 2021).

It should be noted that certain kinds of inorganic nanoparticles are already used commercially by the agricultural industry. For instance, the Center for Food Safety (www.centerforfoodsafety.org), a non-profit organization based in the US, reports several different commercial pesticides that contain inorganic nanoparticles, including those comprised of silver, copper, and titanium dioxide.

3.2.2 Organic nanopesticides

Due to the potential health and environmental risks that have been linked with some kinds of inorganic nanoparticles, there has been interest in creating organic nanopesticides fabricated from natural ingredients, especially those derived from plants (Pavela and Benelli 2016, Pradhan and Mailapalli 2020). Plants produce a diversity of secondary metabolites, many of which are designed to protect them from pests, such as microbes, insects, or herbivores. These natural substances do this through several biological and physicochemical mechanisms, including antimicrobial activity, bitter taste, unpleasant aroma, or exhibiting toxicity to specific species (Wink 2022). Secondary metabolites can be extracted from plants and then used as natural pesticides to help prevent pests from contaminating and destroying crops. Essential oils are one of the most common plant-derived extracts that have been used as a natural pesticide (Pavela and Benelli 2016, Rao et al. 2019). The handling and effectiveness of essential oils is often enhanced by converting them into oil-in-water nanoemulsions, which consist of emulsifier-coated essential oil droplets dispersed within water. These kinds of nanoemulsions are typically low-viscosity fluids that can be sprayed onto plants or soils to provide protection against pests. Other kinds of botanical extracts, including alkaloids, flavonoids, and phenolics, also have good antimicrobial activity and may therefore be used to formulate nanoparticles with pesticide activity (Zaynab et al. 2018). There have been numerous studies on the formulation of nanoenabled delivery systems containing different types of botanical extracts, as well as testing their efficacy against various kinds of pests on different types of agricultural crops (Feng et al. 2018, Pradhan and Mailapalli 2020). Several examples of the utilization of organic nanoparticles as pesticides are given here:

- *Microbes*: Lavender oil nanoemulsions have been shown to be effective against various kinds of microorganisms that contaminate agricultural crops, including gram-negative bacteria (*Salmonella typhimurium*), gram-positive bacteria (*Staphylococcus aureus*), fungi (*Aspergillus flavus* and *Aspergillus niger*), and yeast (*Candida albicans*) (Badr et al. 2021). Clove oil and black seed oil nanoemulsions have been shown to exhibit antifungal effects against various species (*Galactomyces candidum, Alternaria tenuissima,* and *Fusarium solani*) (Mossa et al. 2021). These antifungal nanoemulsions were then shown to inhibit the formation of postharvest rot of cucumber, while not exhibiting any signs of toxicity or mortality in rats at the levels employed.

- *Insects*: Garlic oil nanoemulsions have been shown to be effective at eliminating olive mites, without exhibiting toxicity in a model animal (rats) (Mossa et al. 2018). Various kinds of essential oil nanoemulsions have been shown to be effective at killing aphids that can infect important agricultural crops, such as mung beans (Hashem et al. 2018). Essential oil nanoemulsions have also been shown to inhibit the growth of larvae on wheat during storage (Kavallieratos et al. 2021).

In the future, it may be possible to reduce the levels of synthetic pesticides used by formulating more efficacious plant-based alternatives, such as the essential oil nanoemulsions discussed here. However, there may also be some safety and environmental concerns in using organic nanoparticles to treat agricultural crops (Schappo et al. 2022), but these are likely to be much less than for inorganic ones. Moreover, it is important to optimize the dose of the nanoemulsions used to treat the crops. If it is too low it will be ineffective, but if it is too high it may damage the crops and reduce their yield and nutritional content (Zhang et al. 2021).

3.2.3 Smart nanopesticide delivery systems

Smart pesticides with controlled or triggered release properties can also be produced to increase their efficacy and reduce the dose required, therefore reducing their tendency to contaminate foods or cause pollution (Pradhan and Mailapalli 2020, Li et al. 2021). Various kinds of nanostructured materials can be used to create these kinds of controlled delivery systems including particles, films, and fibers assembled from different substances including polymers, surfactants, metals, and lipids (Singh et al. 2020). As an example, smart pesticides can be formulated by encapsulating the nanopesticides within hydrogel beads (microgels). The kinetics of pesticide release can then be controlled by altering the properties of the microgels, such as their internal pore size, their external diameter, or their interactions with the nanoparticles containing the pesticides (Figure 3.3). For instance, the release of encapsulated antimicrobial nanoparticles from a microgel tends to be slower when the microgel pore size decreases, the microgel diameter increases, or the attractive interactions between the nanoparticles and microgel matrix become stronger. As a result, it may be possible to achieve burst or prolonged release profiles by manipulating the microgel and nanoparticle properties. A triggered release of nanopesticides from microgels can be achieved by designing microgels that respond to specific changes in environmental conditions, like temperature, relative

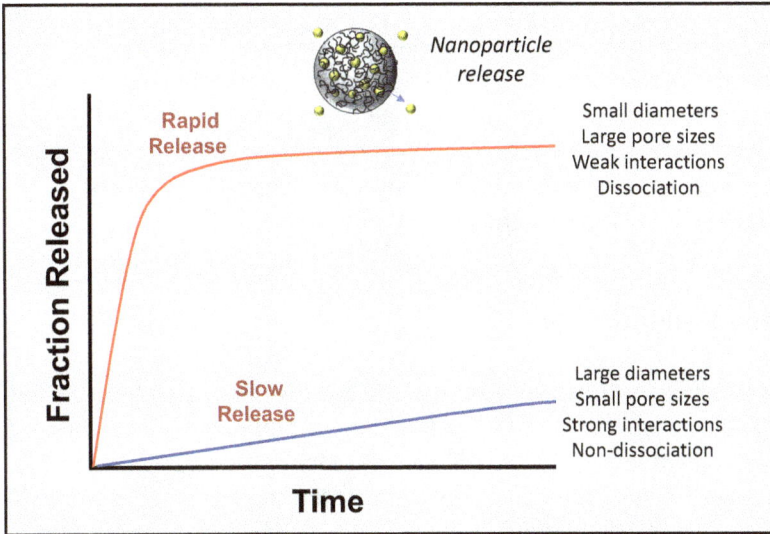

Figure 3.3: The release of nanopesticides or nanofertilizers from microgel particles can be controlled by manipulating their dimensions, pore sizes, interactions, or stability.

humidity, pH, or enzyme activity (Camara et al. 2019). For instance, a microgel could be used that dissociates when it is heated above a critical temperature, thereby leading to a temperature-triggered release of the nanopesticides. This might be important because the pests that damage the crops may only become active when the temperature exceeds a particular level. Some examples of controlled or triggered release systems designed for pesticides are given below:

- Researchers have fabricated pH-responsive nanostructured alginate/chitosan-based microgels to provide a sustained release of a pesticide (imidacloprid) (Singh et al. 2022). These smart delivery systems were shown to be effective against contamination of mung beans by aphids, without damaging the beans. Moreover, they could protect the pesticide from degradation when exposed to UV light.
- Researchers have developed pH-responsive nanostructured alginate-based microgels to control the release of an insecticide (chlorpyrifos) commonly used on agricultural crops (Xiang et al. 2018). These smart nanopesticide delivery systems were then shown to be effective at controlling infestation against grubs without causing damage to plants (foxtail millet).
- Researchers have designed alginate nanogels to encapsulate and control the release of a hydrophilic pesticide (dicamba) (Artusio et al. 2021). These delivery systems were shown to retard the release of the pesticide over a course of days, thereby leading to prolonged treatment. The authors claimed that these nanogels could be used for more effective control of pests on agricultural crops.

In general, various advantages have been claimed for using biopolymer microgels or nanogels to control the release of pesticides (Campos et al. 2015). For instance, the use of these delivery systems reduces the adverse environmental impacts of pesticides by decreasing leaching, volatilization, and degradation. Moreover, they give better long-term control of pests, meaning that lower doses are required, and repeated applications are not necessary. They may also improve soil health by increasing its water holding capacity and permeability.

3.3 Nanoenabled fertilizers

Agricultural crops need nutrients to grow and stay healthy throughout their lifetime (Mikula et al. 2020, Verma et al. 2022). Some of these nutrients may be obtained from their environment, especially the soil they grow in. However, it is usually necessary to provide additional nutrients to ensure that the crops grow efficiently and that they remain healthy throughout their life cycle, thereby increasing crop yield and reducing agricultural waste. For this reason, farmers usually need to administer synthetic or natural fertilizers to their crops. These fertilizers are usually rich in macronutrients (like nitrogen, phosphorous, and potassium) and micronutrients (like boron, chlorine, copper, iron, manganese, molybdenum, and zinc). Ideally, the farmer needs to apply just enough fertilizer to obtain the desired effects. If too much fertilizer is used, then there will be increased costs, as well as increased pollution of the environment (such as eutrophication). Moreover, the fertilizers may need to be applied to the crops at specific times during their life cycle to gain the most benefit. From an economic viewpoint, the lower the number of applications of the fertilizers required the better, since this reduces agricultural production costs.

It has been reported that around 187 million tons per year of fertilizer are currently used globally to produce agricultural crops (Usman et al. 2020). But as much as three quarters of these fertilizers may be lost after being applied to the crops as a result of volatilization, leaching, or run-off (Velez et al. 2021). The loss of these fertilizers is a problem because it can be damaging to the health of both humans and the environment. Consequently, there is a need to identify effective approaches to reduce the amounts of fertilizers that are used and lost.

Nanotechnology approaches like those used to increase the efficacy of pesticides may also be used to improve the effectiveness of fertilizers (Monreal et al. 2016, Mikula et al. 2020, Verma et al. 2022). For instance, fertilizers can be encapsulated within nanoparticles, which are then administered to the crops in a powdered or fluid form. The performance of the fertilizers can be modulated by controlling the composition, size, shape, or interfacial characteristics of the nanoparticles containing them (Figure 3.1). Research has demonstrated that nanofertilizers can be designed to have better efficacy than conventional fertilizers because of the smaller particle size and larger specific surface area (Ma et al. 2018, Adisa et al. 2019, Fatima et al. 2021, Toksha et al. 2021). Indeed, well-designed nanofertilizer formulations have been shown to be better at promoting the growth, decreasing the loss, and improving the nutritional profiles of agricultural crops (Servin et al. 2015). On the other hand, some nanofertilizer formulations have been reported to exhibit the opposite effects for certain kinds of agricultural crops, which highlights the importance of optimizing their properties for each application.

The potential efficacy of well-designed nanofertilizers has been demonstrated using an important agricultural crop: mung beans (Raliya et al. 2016). The researchers used a biosynthesis method to create ZnO nanoparticles that were designed to increase the phosphorous uptake by plants, since zinc acts as a cofactor for phosphorous-solubilizing enzymes. The nanoparticles were shown to increase the activity of these enzymes, which led to an increase in phosphorous uptake, resulting in improvements in stem height, root volume, and biochemical indicators of plant health. Analysis of the plants using mass spectrometry showed that the ZnO nanoparticles had been internalized by them and distributed to all regions of the plants. Even so, the zinc concentration within the edible portions of the plants was still within the values given in dietary recommendations.

Several mechanisms of action have been proposed to account for the improved efficacy of nanofertilizers (Ma et al. 2018):

(1) The small size of nanoparticles means they can penetrate through the pores in the leaves, roots, and stems of agricultural crops, which means they can accumulate inside the plants where nutrients are needed.

(2) The interfacial properties of the nanoparticles used to encapsulate fertilizers can be designed to increase their ability to strongly adhere to the surfaces of plants. As a result, the nanofertilizer tends to remain attached to the plant, meaning that less migrates into and pollutes the environment.

(3) Several kinds of fertilizers and/or pesticides may be included within a single formulation. This can be achieved by incorporating these substances into one type of nanoparticle or by using multiple types of nanoparticles with different compositions.

(4) Some nanofertilizers can deactivate pests that would normally inhibit plant growth, thereby promoting the health and yield of agricultural crops.

(5) Nanofertilizers can improve crop nutrition, thereby enhancing the defense mechanisms of the plant, which again promotes their health and yield (Adisa et al. 2019).

In a recent study, it was shown that functionalized carbon-based nanoparticles could be used to enhance the photosynthesis of plants (Swift et al. 2021). The functionalized nanoparticles were small enough to be taken up by the plants through the soil. They then provided increased photoprotection of the plants by stimulating the production of carotenoids. Moreover, they increased the efficiency of photosynthesis by the plant (wheat), which led to an increase in around 18% of the grain yield, without reducing grain quality. The nanoparticles used consisted of carbon nanodots that were functionalized by attaching glucose units to their surfaces, which increased their uptake by the plants, as well as their efficacy. The authors claimed that these nanoparticles were safe to the environment and humans. These kinds of carbon nanodots may therefore act as a novel kind of nanofertilizer for agricultural crops.

If they are to be commercially viable, it will be important to show that nanofertilizers and nanopesticides are effective, safe, environmentally friendly, easy to use, and cost-effective. Conventional fertilizers and pesticides are typically inexpensive commodities that are used around the world in large amounts. Much of the research on encapsulating fertilizers and pesticides within nanoparticles has been performed in academic laboratories using ingredients and processes that may be too expensive and/or too challenging to scale-up. Having said this, several kinds of nanoparticles can be produced economically on a large scale, e.g., nano-copper, nano-silver, and nanoemulsions. Indeed, some of these are already commercially available (see end of this section). Moreover, the increased costs of using nanofertilizers or nanopesticides are often offset by the savings arising from the increase in crop yields and reduction in crop losses. For instance, researchers have reported that treating watermelon with a nanofertilizer cost by about $26 per acre but led to an increase in yield that was worth about $4,600 per acre (Dimkpa and Bindraban 2017). Thus, the increased costs associated with using the new nanofertilizer formulation were more than offset by the increased yields it led to.

The various steps needed to develop and employ nanofertilizers and nanopesticides that are safe, effective, and commercially viable have been outlined elsewhere (Rodrigues et al. 2017). Detailed life-cycle analyses are required to determine the transfer route of the nanoparticles from soils to plants to insects to animals to humans and to the environment. Moreover, an increased understanding of the potentially harmful impacts of nanoparticles on agricultural crops, people, and the environment is required. For instance, more knowledge is required about the persistence of nanoparticles within the environment, as well as to determine their influence on plant and soil microbiomes.

The Center for Food Safety (www.centerforfoodsafety.org), a non-profit organization based in the US, reports several different commercial fertilizers that contain nanoparticles, including those containing zinc, phosphate, calcium, and copper.

3.4 Nanoenabled aquaculture and animal feed

Aquaculture is making an increasingly large contribution to the global seafood supply (Naylor et al. 2021). This increase has been largely driven by the need to overcome problems associated with overfishing, pollution, and changes in the migratory pattern of open water fish stocks due to climate change. In aquaculture, the fish are raised in contained areas, which may be located, on land, lakes, or oceans. There are several ways in which nanotechnology can improve the efficiency of aquaculture (Shah and Mraz 2020, Nasr-Eldahan et al. 2021, Jeyavani et al. 2022).

Fish preservation treatments: A major concern in the modern aquaculture industry is the infection of the fish stock with pathogenic microorganisms or other pests (Jeyavani et al. 2022). These pathogens can cause the fish to become sick, thereby reducing yields, adversely altering nutritional profiles, and leading to food safety problems. Consequently, there is a need to develop effective strategies to prevent or treat the contamination of aquaculture fish stocks from pests. Traditionally, prophylactic chemical-based therapeutics, antibiotics, and vaccines have been utilized to increase the resilience of fish stocks to diseases. In addition, antimicrobial agents can be used to treat prevent contamination of the fish or the waters they inhabit. For instance, these agents may have antiprotozoal, antifungal, antiviral, and/or antibacterial properties. However, the widespread utilization of conventional prophylactics and pesticides can cause pollution of the environment, contamination of the foods produced from fish, and the development of antibiotic resistance. Consequently, researchers have been developing alternative strategies for ensuring the fish stock remains healthy, without causing undesirable environmental or human health concerns. For instance, there has been great interest in the utilization of antimicrobial phytochemicals (such as curcumin, quercetin, silymarin, caffeic acid, and essential oils) to increase the immunity of fish and to prevent or treat fish diseases (Jeyavani et al. 2022). However, these phytochemicals are often strongly hydrophobic substances that have a very low water solubility and are therefore difficult to administer to fish. This problem can be overcome by encapsulating these molecules in colloidal dispersions containing lipid nanoparticles, such as nanoemulsions. These phytochemical-loaded colloidal dispersions can then be introduced into the fishes' water, or they can be further encapsulated in small beads that can be used as fish feed that is added to the water (Figure 3.4). The size, composition, and interfacial properties of the lipid nanoparticles can be controlled to increase the dispersibility, stability, and bioavailability of the encapsulated phytochemicals. These phytochemical-loaded nanoemulsions are therefore part of the growing interest in the development of "organic aquaculture" systems.

Water purification: Nanoenabled preservatives can also be utilized to sterilize the water that the fish reside in by increasing the efficacy of antimicrobial agents (Shah and Mraz 2020). In this case, metal nanoparticles, like silver or gold ones, are often used because of their strong antimicrobial activity but there are concerns about their potential for polluting the environment and contaminating the foods obtained from the fish. Alternatively, nanostructured hydrogels may be utilized as absorbents to remove ionic species or heavy metals that may contaminate the water, such as arsenic, nitrates, fluorides, lead, or mercury. These hydrogels contain substances that can pull these harmful substances from the water and then retain them.

Feed for fish: Captive fish must have adequate nutrition throughout their life cycle to ensure they remain healthy and grow to the required size. Normally, fish would meet their nutritional needs in their natural environments. In aquaculture, however, the fish must get their feed administered to them. Nanotechnology can be used to increase the potency of the fish feed used in aquaculture. As an example, nutrients and nutraceuticals can be converted into nanoparticles that are easier to administer to the fish and that are rapidly digested and absorbed after ingestion (Shah and Mraz 2020). The bioavailability of essential vitamins and minerals can often be improved by incorporating them into nanosized particles, which are then fed to the fish (Figure 3.4). These nanoforms of nutrients and nutraceuticals may then have beneficial effects on fish health and yields.

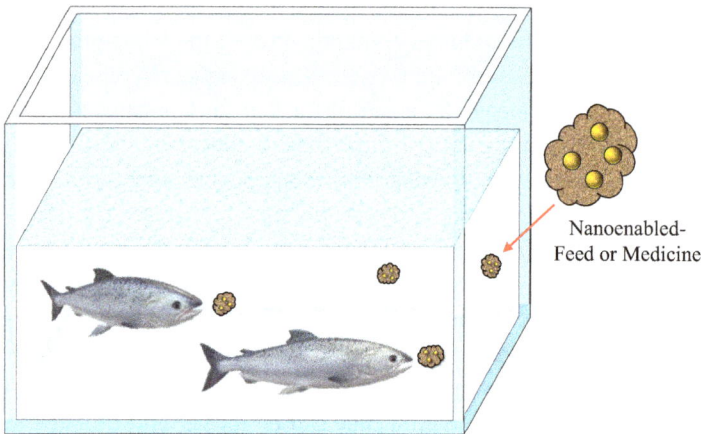

Nanoenabled-
Feed or Medicine

Figure 3.4: Encapsulation of bioactive ingredients in nanoparticles can be used to improve the dispersibility, stability, and bioavailability of the nutrients and medicines for animals, such as the fish in aquaculture.

Feed for other animals: Nanomaterials have also been incorporated into the feed of various kinds of livestock animals to improve their nutrition or health (Peters et al. 2016). For instance, converting vitamins, minerals, or nutraceuticals into nanosized particles can increase their bioavailability since they tend to be digested, solubilized, and absorbed more rapidly in the animal's gastrointestinal tract than their conventional forms. Incorporating silver nanoparticles into the water used to feed chickens has been reported to have an effect analogous to conventional antibiotics, which is attributed to the antimicrobial activity of this kind of nanoparticle.

3.5 Nano-assisted precision agriculture

Precision agriculture can be utilized to increase yields, reduce losses, and improve the nutritional quality of agricultural crops (Monisha and Dhanalakshmi 2015, Bhakta

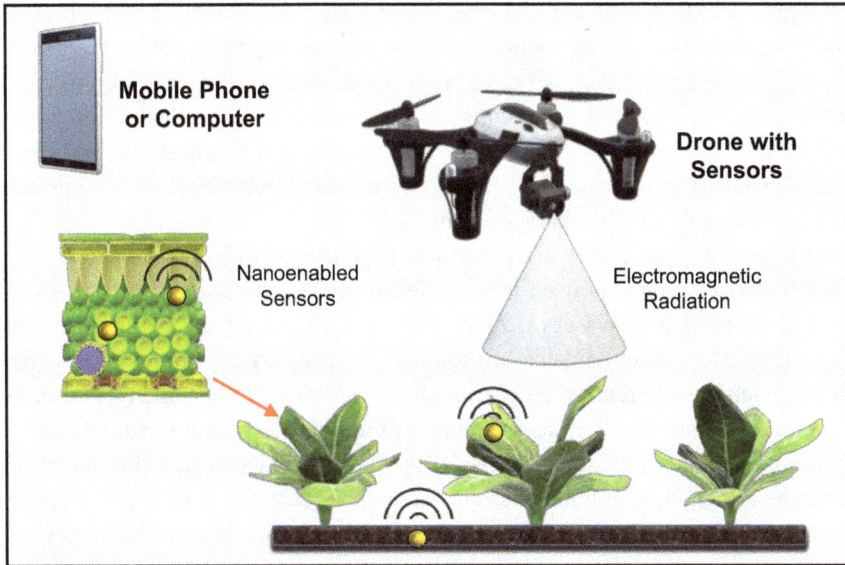

Figure 3.5: Nanoparticles internalized into plant cells or soil may be used as sensors to provide information about the maturity, health status, or water/nutrient needs of agricultural crops.

et al. 2019, Shafi et al. 2019, Monteiro et al. 2021). This involves using a combination of advanced technologies, including sensors, automation, and artificial intelligence, to provide more detailed information about agricultural crops throughout their life span, and then act on this information (Shafi et al. 2019, Saleem et al. 2021). For instance, these technologies can be used to monitor the growth stage, health status, nutritional requirements, water needs, and contamination of agricultural crops throughout their life cycle, as well as changes in the properties of the soil they grow in and the climatic conditions they experience (such as sunlight, temperature, humidity, wind, and rainfall). The data is then collected, stored, and analyzed and appropriate responses are made, such as treating the crops with appropriate types and amounts of water, fertilizers, and/or pesticides (McClements 2019). This can be achieved using automated machines (such as robotic tractors or drones) that apply these substances to the plants precisely when they need them (Figure 3.5). Moreover, they can be applied only to the plants that need them, or the ratio of water, fertilizers, and pesticides can be customized for each type of plant. Advanced sensor technologies can also be used to establish just the right time to harvest the crops, which can then be carried out using other types of automated machines (such as robotic harvesters). The effective performance of these integrated automated systems depends on having precise and detailed information about the crops, soil, and climatic conditions, which requires the availability of advances sensor technologies (Wang et al. 2016, Thakur et al. 2019, Chapungo et al. 2021). Advanced spectroscopy methods such as multispectral imaging are being utilized for this purpose (Hassan et al. 2021), but nanotechnology may also play an important role in the development of these sensors (Srivastava et al. 2017). As discussed in the previous section, nanotechnology can also be used to create a new generation of pesticides and fertilizers, which can also be integrated into precision agricultural practices. For instance, they could be incorporated into a fluid form that can be sprayed precisely onto the surfaces of crops using drones or robotic tractors.

Nanotechnology has been widely explored as a method to increase the sensitivity and selectivity of sensors used to provide information about the properties of agricultural crops (Srivastava et al. 2017). For instance, nanoparticles have a very high specific surface area because of their small dimensions. Ligands can be attached to the surfaces of these nanoparticles that will bind to a specific type of molecule within a test sample, such as a biomarker for a plant's health status or degree of contamination. The high surface area of the nanoparticles means that many ligands can be attached, which increases the sensitivity of the device. When a ligand binds, the system is designed to give a measurable change that can be observed and/or recorded, such as a change in color or electrical resistance.

The development and application of nanobased sensor technologies for application in food and agriculture are discussed in more detail in a later chapter. For this reason, only a brief overview is given here. Typically, the sample to be tested, for instance a plant tissue or soil sample, is collected, ground, dispersed in water, and then placed within the measurement chamber of the device. The device contains a sensor that gives a signal that can be detected and recorded (such as a change in color or electrical resistance) when a specific target substance contacts it. The information provided by this device can then be used to rapidly decide how the crops should be treated (e.g., watered, treated with pesticides, treated with fertilizers, or cultivated).

Nanoenabled sensor technologies are being developed that can be situated inside the crops or the soil they grow in, which can send information to a remote device, such as a mobile phone or computer (Atzberger 2013). These types of sensors could provide detailed information about the maturity, nutritional requirements, and health status of plants throughout their lifespan, thereby providing much better control of agricultural production. When designing sensors that are located within crops or soils it is critical that they are affordable, robust, dependable, easy to use, and safe. As an example of this approach, researchers have shown that sensors based on carbon nanotubes can be localized inside plants, where they can then transmit signals to a smartphone about whether they have been contaminated by pesticides or not (Wong et al. 2017). It has been reported that bioelectronic devices suitable for incorporation inside plant tissues can be affordably mass produced using organic or carbon-based materials, which may lead to their more widespread utilization (Stavrinidou et al. 2022). If these sensors remain inside the edible parts of crops, it will be important to ensure they do not exhibit any toxicity to humans or livestock animals fed the crops. Alternatively, a small number of plants may be selected and then their health and nutritional status is monitored using the internalized nanosensors. The information obtained from these representative plants is then used to assess the behavior of the whole crop, without needing to consume them. The data obtained from these sensors could be stored and analyzed using artificial intelligence to obtain relationships between environmental conditions, soil quality, and plant health, which may lead to more efficient agricultural processes.

3.6 Valorizing food and agricultural waste streams

Nanotechnology can also be used to improve the economic viability and sustainability of the food supply by valorizing waste streams produced by the food and agricultural industries (Roy et al. 2021). A considerable fraction of an agricultural crop may not be used for food. If it cannot be used for another purpose (such as animal feed), then it may be discarded, which leads to an increase in waste and pollution. Food processing factories often produce effluents that are rich in nutrients and other potentially useful functional components (such as colors, flavors, antioxidants, antimicrobials, nutraceuticals, and dietary fibers), but which are currently discarded. Again, these effluents may contaminate the environment or lead to increased waste. There is therefore great interest in developing effective processes to convert these waste streams into value-added products.

3.6.1 Advanced separation processes

Nutrients and other functional components can often be isolated from waste streams using separation-based processing operations, such as centrifugation, gravitational settling, or filtration (Bacchin et al. 2011). The principles of nanotechnology can often be used to enhance the efficiency of these processes. As an example, knowledge of the colloidal interactions operating among particles or macromolecules, which typically occur on length scales in the nanometer range (1 to 50 nm), can be used to control their state of aggregation (individual particles or agglomerates), which is important when trying to remove them from solution. For instance, conditions can be altered to promote the attraction between the particles or macromolecules, which causes them to aggregate and precipitate, thereby facilitating their removal by filtration, centrifugation, of gravitational separation methods. Selective control of the colloidal interactions between different kinds of substances in complex mixtures can be used to selectively precipitate specific components, such as proteins, starch, dietary fibers, lipids, and polyphenols.

Nanotechnology has been shown to be capable of creating novel filters that can selectively extract certain substances from waste streams (Ahmed et al. 2014, Bhagat et al. 2015, McClements 2019). As an example, nanofilters have been fabricated by electrospinning synthetic or natural polymers to create porous materials consisting of a network of highly entangled polymer nanofibers (Soares et al. 2018). The surface chemistry, dimensions, pore size, and surface area of these nanofilters can be manipulated so they can efficiently and selectively remove specific types of substances from complex mixtures. These types of nanofilters have been utilized to purify water, as well as to remove valuable substances from waste streams. For example, carbon nanofibers have been shown to be effective at removing chlorinated pesticides from ginseng extracts, which is useful for reducing their potential toxicity (Sun et al. 2021). Chitosan-based nanofibers produced by electrospinning have been shown to exhibit antimicrobial properties and to be effective filters, e.g., for removing heavy metals from water (Sun et al. 2021). Similarly, composite nanofibers fabricated from polycaprolactone and cellulose using electrospinning have been shown to be effective at removing heavy metals from water (Hinestroza et al. 2020). A recent study showed that composite nanomaterials comprised of activated carbon and milk protein nanofibers could be used to remove arsenic from contaminated water (Bolisetty et al. 2021). The same research group showed that similar systems could be used to remove radionuclides (technetium, iodine, gallium, iodine, and lutetium) and heavy metals (chromium, nickel, silver, and platinum) from contaminated water (Peydayesh et al. 2019, Bolisetty et al. 2020). The protein nanofibers were created by heating globular milk proteins above their thermal denaturation temperature under acidic conditions, which promoted the formation of long strings of protein molecules (amyloid fibrils). The diverse surface chemistry of these protein nanofibers, which is a result of the many different amino acid side groups they contain, means they can adsorb a wide range of contaminants (Peydayesh et al. 2019). Nanofibrous poly(ethylene terephthalate) membranes produced by electrospinning have been used for the clarification of fruit juices (Veleirinho and Lopes-da-Silva 2009). These nanoenabled membranes were also shown to have good mechanical properties, allow high throughputs to be achieved, and were simple to use. Taken together, these studies suggest that nanotechnology may have some important applications in improving the efficacy of separation technologies in the food and agricultural industry.

3.6.2 Manufacture of nano-enabled food ingredients

In some cases, nanomaterials with desirable functional attributes can be extracted from substances produced by the food or agricultural industries. Various kinds of cellulose-based nanomaterials can be obtained from wood, cotton, and other waste material, including cellulose nanofibrils (CNF) and cellulose nanocrystals (CNC). These nanomaterials are usually isolated by partial dissociation and disintegration of the cellulose fibers found in these materials using a combination of chemical and mechanical treatments (Bai et al. 2021). Cellulose nanomaterials can also be obtained from some microbial sources, such as bacterial nanocellulose (BNC) (Montoya et al. 2019). CNC, CNF, and BCN differ in their dimensions, shapes, and surface chemistries because of differences in their origin and the processing operations used to extract and purify them, which leads to differences in their functional performances in different applications. Other kinds of nanomaterials can be isolated from natural sources, such as nanochitin from the shells of crustaceans (Jeevahan and Chandrasekaran 2019, Joseph et al. 2020). This material has novel functional attributes because of the positive charge associated with its amino surface groups.

Some of the potential applications of nanocelluloses as functional ingredients are briefly highlighted here (Bai et al. 2021).

- *Particle-based emulsifiers*: Various kinds of nanocelluloses are efficient particle-based emulsifiers capable of stabilizing Pickering emulsions. Cellulose-based nanoparticles or nanofibers can adsorb to oil-water interfaces and form a protective coating around the oil droplets, which inhibits their tendency to coalesce with each other. As a result, they can form oil-in-water emulsions that are highly resistance to phase separation during storage. Indeed, emulsions stabilized by these kinds of nanoparticles typically have much better stability to coalescence than emulsions stabilized by small molecules.
- *Modulation of gastrointestinal fate of nutrients*: Nanocellulose can be added to food and beverage products to inhibit the digestion of macronutrients, such as lipids, starches, and proteins. The presence of the nanocellulose can inhibit digestion through several mechanisms, including increasing the viscosity of the gastrointestinal fluids, slowing down mixing and diffusion processes, forming protective coatings around macronutrients, and binding to gastrointestinal constituents like digestive enzymes, bile salts, or calcium (Figure 3.6). The ability to slow down macronutrient digestion may be beneficial to human health, e.g., by preventing the spikes in blood sugar or lipid levels that may occur after consumption of foods containing high levels of starch or fat (DeLoid et al. 2018). As a result, these kinds of nanocellulose-enriched functional foods could help to combat overeating and diabetes. However, the presence of nanocellulose may reduce the bioavailability of bioactive components, such as oil soluble vitamins and nutraceuticals (Zhou et al. 2021, Fitri et al. 2022). This is because inhibition of lipid digestion will mean that less of these hydrophobic bioactives are released from the oil phase and solubilized in the mixed micelles (Figure 3.6). Other kinds of nanomaterials isolated from waste streams from the food or agricultural industries have also been shown to have similar effects, such as nanochitin, which can be obtained from crab shells (Zhou et al. 2020).
- *Fat replacers*: Nanocellulose can also be used as a fat replacer to reduce the calorie content of fatty foods, such as mayonnaise, sauces, dressings, and meat products (Heggset et al. 2020, Marchetti and Andres 2021). They do this because they are small water-insoluble particles that mimic some of the desirable optical, textural, and sensory attributes normally provided by the fat droplets in

foods. Researchers have shown that nanocellulose can be used as a fat replacer in meat products, without adversely affecting their texture and flavor profile (Qi et al. 2020, Marchetti and Andres 2021). Nanocellulose has also been used to improve the textural attributes of low-fat ice products (Velasquez-Cock et al. 2019). Finally, it has been used to replace fat and starch in model reduced-calorie mayonnaise products, where it could provide the textural attributes normally provided by these components (Heggset et al. 2020).

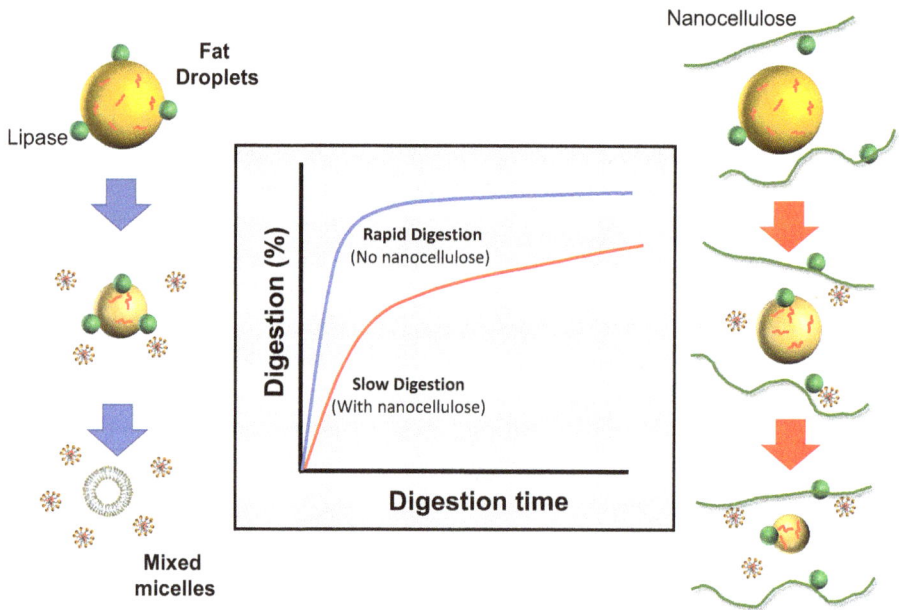

Figure 3.6: The presence of nanocellulose in foods may inhibit the digestion of lipids within the human gastrointestinal tract, which may also reduce the bioaccessibility of oil-soluble vitamins and nutraceuticals.

3.6.3 Manufacture of biodegradable packaging materials

Nanomaterials may also be used as structure forming or functional ingredients in next-generation smart or active packaging materials. Recently, there has been a strong focus on their utilization in the production of biodegradable packaging materials as sustainable alternatives to petroleum-based plastics (Vilarinho et al. 2018, Fotie et al. 2020, Ahankari et al. 2021). This kind of packaging material is often produced from biopolymers such as proteins and polysaccharides. However, these kinds of biodegradable packaging materials often do not have the optical, barrier, or mechanical properties required for commercial applications. The incorporation of nanomaterials, like nanocellulose, nanochitin, or nano-clay, can be used to increase the mechanical strength, light screening, and barrier properties of biopolymer-based films and coatings. Moreover, they can be used to create active packaging materials with antioxidant and antimicrobial features that help preserve foods for longer, or smart packaging materials with sensing capabilities that can provide information about food quality, freshness, or safety (Vigneshwaran et al. 2019, Sani et al. 2021). A more detailed discussion of the properties of nano-structured packaging materials is given in Chapter 5.

3.6.4 Miscellaneous applications

Nanomaterials derived from natural resources can also be used for various non-food applications. For instance, dehydrated cellulose nanofibers are capable of absorbing large quantities of liquids (including blood and urine), which makes them useful in the creation of more environmentally friendly tampons, diapers, and incontinence pads (McClements 2019). Biodegradable food serving containers have been fabricated from cellulose nanofibrils or lignin-containing cellulose nanofibrils isolated from wood waste (Hossain et al. 2021). These containers were shown to have resistance to oil and grease, strong mechanical properties, and good thermal stability. Consequently, they could be used as replacements for conventional petroleum-based serving containers, such as plastic cups and plates.

Another functional nanomaterial that can be produced from agricultural waste products is biochar (Barrow 2012). Biochar is created by burning agricultural waste in the presence of relatively low levels of oxygen, which leads to the production of a powder containing high concentrations of carbon and minerals. This powder can then be used in agriculture as a fertilizer or to improve soil health (Agegnehu et al. 2017, Chen et al. 2019). It can also be used to treat wastewater by removing toxic metals, organic pollutants, and nutrients, which has been attributed to its large specific surface area and strong adsorption capacity (Xiang et al. 2020). An additional benefit to producing biochar is that it can help to decrease greenhouse gas emissions, especially carbon dioxide, thereby helping to combat global warming.

3.7 Conclusions

Researchers have identified a diverse range of potential applications of nanomaterials (especially nanoparticles and nanofibers) within agriculture. Nanopesticides and nanofertilizers have already been shown to have some advantages over conventional pesticides and fertilizers that have led them to be employed by the industry to improve crop yields, increase nutritional value, enhance resilience, and reduce waste. The utilization of these advanced nanomaterials may therefore be able to enhance the efficiency and sustainability of modern agriculture. Nanotechnology can also be utilized for various other applications, such as to create animal and fish feeds with enhanced nutritional specifications or that can improve their health by combating pathogenic microorganisms or other pests. Moreover, nanotechnology principles are being used to convert some of the side streams and waste products of the agricultural industry into value-added functional materials, such as biodegradable packaging, nanofilters, and novel food ingredients. Nevertheless, it is critical to ensure that any nanomaterials used in the agricultural industry do not have adverse effects on human or environmental health. Consequently, systematic investigations into the fate of nanomaterials within soils, plants, insects, animals, humans, and the environment are needed, so that their potential to cause toxicity and pollution can be understood and controlled (Deng et al. 2014, Kurwadkar et al. 2015, Ali et al. 2021, Bakshi and Kumar 2021). A more detailed discussion of the potential toxicity of nanoparticles is given in Chapter 6.

References

Adisa, I. O., V. L. R. Pullagurala, J. R. Peralta-Videa, C. O. Dimkpa, W. H. Elmer, J. L. Gardea-Torresdey and J. C. White (2019). "Recent advances in nano-enabled fertilizers and pesticides: a critical review of mechanisms of action". Environmental Science-Nano 6(7): 2002–2030.

Agegnehu, G., A. K. Srivastava and M. I. Bird (2017). "The role of biochar and biochar-compost in improving soil quality and crop performance: A review". Applied Soil Ecology 119: 156–170.

Ahankari, S. S., A. R. Subhedar, S. S. Bhadauria and A. Dufresne (2021). "Nanocellulose in food packaging: A review". Carbohydrate Polymers 255: 1–15. doi.org/10.1016/j.carbpol.2020.117479

Ahmed, T., S. Imdad, K. Yaldram, N. M. Butt and A. Pervez (2014). "Emerging nanotechnology-based methods for water purification: a review". Desalination and Water Treatment 52 (22–24): 4089–4101.

Ali, S., A. Mehmood and N. Khan (2021). "Uptake, translocation, and consequences of nanomaterials on plant growth and stress adaptation". Journal of Nanomaterials 2021: 1–17. doi.org/10.1155/2021/6677616

Artusio, F., D. Casa, M. Granetto, T. Tosco and R. Pisano (2021). "Alginate nanohydrogels as a biocompatible platform for the controlled release of a hydrophilic herbicide". Processes 9(9): 1641, 1–17. doi.org/10.3390/pr9091641

Atzberger, C. (2013). "Advances in remote sensing of agriculture: context description, existing operational monitoring systems and major information needs". Remote Sensing 5(2): 949–981.

Bacchin, P., A. Marty, P. Duru, M. Meireles and P. Aimar (2011). "Colloidal surface interactions and membrane fouling: Investigations at pore scale". Advances in Colloid and Interface Science 164(1–2): 2–11.

Badr, M. M., M. E. I. Badawy and N. E. M. Taktak (2021). "Characterization, antimicrobial activity, and antioxidant activity of the nanoemulsions of Lavandula spica essential oil and its main monoterpenes". Journal of Drug Delivery Science and Technology 65: 1–12. doi.org/10.1016/j.jddst.2021.102732

Bai, L., S. Q. Huan, Y. Zhu, G. Chu, D. J. McClements and O. J. Rojas (2021). "Recent advances in food emulsions and engineering foodstuffs using plant-based nanocelluloses". Annual Review of Food Science and Technology Vol 12, 2021. M. Doyle and D. J. McClements **12**: 383–406.

Bakshi, M. and A. Kumar (2021). "Copper-based nanoparticles in the soil-plant environment: Assessing their applications, interactions, fate and toxicity". Chemosphere **281**: 1–14. doi.org/10.1016/j.chemosphere.2021.130940

Barrow, C. J. (2012). "Biochar: Potential for countering land degradation and for improving agriculture". Applied Geography **34**: 21–28.

Bhagat, Y., K. Gangadhara, C. Rabinal, G. Chaudhari and P. Ugale (2015). "Nanotechnology in agriculture: A review". Journal of Pure and Applied Microbiology **9**(1): 737–747.

Bhakta, I., S. Phadikar and K. Majumder (2019). "State-of-the-art technologies in precision agriculture: a systematic review". Journal of the Science of Food and Agriculture **99**(11): 4878–4888.

Bolisetty, S., N. M. Coray, A. Palika, G. A. Prenosil and R. Mezzenga (2020). "Amyloid hybrid membranes for removal of clinical and nuclear radioactive wastewater". Environmental Science-Water Research & Technology **6**(12): 3249–3254.

Bolisetty, S., A. Rahimi and R. Mezzenga (2021). "Arsenic removal from Peruvian drinking water using milk protein nanofibril-carbon filters: a field study". Environmental Science-Water Research & Technology **7**(12): 2223–2230.

Camara, M. C., E. V. R. Campos, R. A. Monteiro, A. D. S. Pereira, P. L. D. Proenca and L. F. Fraceto (2019). "Development of stimuli-responsive nano-based pesticides: emerging opportunities for agriculture". Journal of Nanobiotechnology **17**(1): 1–19. doi.org/10.1186/s12951-019-0533-8

Campos, E. V. R., J. L. de Oliveira, L. F. Fraceto and B. Singh (2015). "Polysaccharides as safer release systems for agrochemicals". Agronomy for Sustainable Development **35**(1): 47–66.

Chapungo, N. J., O. Postolache and Ieee (2021). Sensors and comunication protocols for precision agriculture. 12th International Symposium on Advanced Topics in Electrical Engineering (ATEE), Bucharest, ROMANIA.

Chaud, M., E. B. Souto, A. Zielinska, P. Severino, F. Batain, J. Oliveira and T. Alves (2021). "Nanopesticides in agriculture: Benefits and challenge in agricultural productivity, toxicological risks to human health and environment". Toxics **9**(6): 1–19. doi.org/10.3390/toxics9060131

Chen, J. N., S. L. Li, J. X. Luo, R. S. Wang and W. Ding (2016). "Enhancement of the antibacterial activity of silver nanoparticles against phytopathogenic bacterium ralstonia solanacearum by stabilization". Journal of Nanomaterials **2016**: 1–15. doi.org/10.1155/2016/7135852

Chen, W. F., J. Meng, X. R. Han, Y. Lan and W. M. Zhang (2019). "Past, present, and future of biochar". Biochar **1**(1): 75–87.

Damalas, C. A. and I. G. Eleftherohorinos (2011). "Pesticide exposure, safety issues, and risk assessment indicators". International Journal of Environmental Research and Public Health **8**(5): 1402–1419.

DeLoid, G. M., I. S. Sohal, L. R. Lorente, R. M. Molina, G. Pyrgiotakis, A. Stevanovic, R. Zhang, D. J. McClements, N. K. Geitner, D. W. Bousfield, K. W. Ng, S. C. J. Loo, D. C. Bell, J. Brain and P. Demokritou (2018). "Reducing intestinal digestion and absorption of fat using a nature-derived biopolymer: interference of triglyceride hydrolysis by nanocellulose". ACS nano **12**(7): 6469–6479.

Deng, Y. Q., J. C. White and B. S. Xing (2014). "Interactions between engineered nanomaterials and agricultural crops: implications for food safety". Journal of Zhejiang University-Science A **15**(8): 552–572.

Dimkpa, C. O. and P. S. Bindraban (2017). "Nanofertilizers: new products for the industry?". Journal of Agricultural and Food Chemistry.**66**(26), 6462–6473.

Elmer, W., R. De La Torre-Roche, L. Pagano, S. Majumdar, N. Zuverza-Mena, C. Dimkpa, J. Gardea-Torresdey and J. C. White (2018). "Effect of metalloid and metal oxide nanoparticles on fusarium wilt of watermelon". Plant Disease **102**(7): 1394–1401.

Fatima, F., A. Hashim and S. Anees (2021). "Efficacy of nanoparticles as nanofertilizer production: a review". Environmental Science and Pollution Research **28**(2): 1292–1303.

Feng, J. G., Q. Zhang, Q. Liu, Z. X. Zhu, D. J. McClements and S. M. Jafari (2018). Application of Nanoemulsions in Formulation of Pesticides.

Fitri, I. A., W. Mitbumrung, P. Akanitkul, N. Rungraung, V. Kemsawasd, S. Jain and T. Winuprasith (2022). "Encapsulation of beta-Carotene in Oil-in-Water Emulsions Containing Nanocellulose: Impact on Emulsion Properties, In Vitro Digestion, and Bioaccessibility". Polymers **14**(7) 1414, 1–17. doi: 10.3390/polym14071414

Fotie, G., S. Limbo and L. Piergiovanni (2020). "Manufacturing of food packaging based on nanocellulose: Current advances and challenges". Nanomaterials **10**(9). 1726: 1–26. https://doi.org/10.3390/nano10091726

Girotto, F., L. Alibardi and R. Cossu (2015). "Food waste generation and industrial uses: A review". Waste Management **45**: 32–41.

Grillo, R., P. C. Abhilash and L. F. Fraceto (2016). "Nanotechnology applied to bio-encapsulation of pesticides". Journal of Nanoscience and Nanotechnology **16**(1): 1231–1234.

Grillo, R., L. F. Fraceto, M. J. B. Amorim, J. J. Scott-Fordsmand, R. Schoonjans and Q. Chaudhry (2021). "Ecotoxicological and regulatory aspects of environmental sustainability of nanopesticides". Journal of Hazardous Materials **404**. 124148: 1–17.

Hashem, A. S., S. S. Awadalla, G. M. Zayed, F. Maggi and G. Benelli (2018). "Pimpinella anisum essential oil nanoemulsions against Tribolium castaneum-insecticidal activity and mode of action". Environmental Science and Pollution Research **25**(19): 18802–18812.

Hassan, S. I., M. M. Alam, U. Illahi, M. A. Al Ghamdi, S. H. Almotiri and M. M. Su'ud (2021). "A systematic review on monitoring and advanced control strategies in smart agriculture". Ieee Access **9**: 32517–32548.

He, X. J., H. Deng and H. M. Hwang (2019). "The current application of nanotechnology in food and agriculture". Journal of Food and Drug Analysis **27**(1): 1–21.

Heggset, E. B., R. Aaen, T. Veslum, M. Henriksson, S. Simon and K. Syverud (2020). "Cellulose nanofibrils as rheology modifier in mayonnaise – A pilot scale demonstration". Food Hydrocolloids **108**:106084: 1–15.

Hinestroza, H. P., H. Urena-Saborio, F. Zurita, A. A. G. de Leon, G. Sundaram and B. Sulbaran-Rangel (2020). "Nanocellulose and polycaprolactone nanospun composite membranes and their potential for the removal of pollutants from water". Molecules **25**(3): 683, 1–13. doi.org/10.3390/molecules25030683

Hong, J., C. Wang, D. C. Wagner, J. L. Gardea-Torresdey, F. He and C. M. Rico (2021). "Foliar application of nanoparticles: mechanisms of absorption, transfer, and multiple impacts". Environmental Science-Nano **8**(5): 1196–1210.

Hossain, R., M. Tajvidi, D. Bousfield and D. J. Gardner (2021). "Multi-layer oil-resistant food serving containers made using cellulose nanofiber coated wood flour composites". Carbohydrate Polymers **267**: 118221, 1–14. doi.org/10.1016/j.carbpol.2021.118221

Huang, Y. X., W. W. Li, A. S. Minakova, T. Anumol and A. A. Keller (2018). "Quantitative analysis of changes in amino acids levels for cucumber (Cucumis sativus) exposed to nano copper". Nanoimpact **12**: 9–17.

Jeevahan, J. and M. Chandrasekaran (2019). "Nanoedible films for food packaging: a review". Journal of Materials Science **54**(19): 12290–12318.

Jeyavani, J., A. Sibiya, J. Sivakamavalli, M. Divya, E. Preetham, B. Vaseeharan and C. Faggio (2022). "Phytotherapy and combined nanoformulations as a promising disease management in aquaculture: a review". Aquaculture International **30**: 1071–1086.

Joseph, B., R. M. Sam, P. Balakrishnan, H. J. Maria, S. Gopi, T. Volova, S. C. M. Fernandes and S. Thomas (2020). "Extraction of nanochitin from marine resources and fabrication of polymer nanocomposites: Recent advances". Polymers **12**(8): 1664, 1–17. doi.org/10.3390/polym12081664

Kah, M. (2015). "Nanopesticides and nanofertilizers: Emerging contaminants or opportunities for risk mitigation?". Frontiers in Chemistry **3**: 1–6. doi.org/10.3389/fchem.2015.00064

Kah, M., L. J. Johnston, R. S. Kookana, W. Bruce, A. Haase, V. Ritz, J. Dinglasan, S. Doak, H. Garelick and V. Gubala (2021). "Comprehensive framework for human health risk assessment of nanopesticides". Nature Nanotechnology **16**(9): 955–964.

Kah, M., D. Navarro, R. S. Kookana, J. K. Kirby, S. Santra, A. Ozcan and S. Kabiri (2019). "Impact of (nano)formulations on the distribution and wash-off of copper pesticides and fertilisers applied on citrus leaves". Environmental Chemistry **16**(6): 401–410.

Kavallieratos, N. G., E. P. Nika, A. Skourti, N. Ntalli, M. C. Boukouvala, C. T. Ntalaka, F. Maggi, R. Rakotosaona, M. Cespi, D. R. Perinelli, A. Canale, G. Bonacucina and G. Benelli (2021). "Developing a Hazomalania voyronii Essential Oil Nanoemulsion for the Eco-Friendly Management of Tribolium confusum, Tribolium castaneum and Tenebrio molitor Larvae and Adults on Stored Wheat". Molecules **26**(6): 1812, 1–9. doi.org/10.3390/molecules26061812

Kibler, K. M., D. Reinhart, C. Hawkins, A. M. Motlagh and J. Wright (2018). "Food waste and the food-energy-water nexus: A review of food waste management alternatives". Waste Management **74**: 52–62.

Kurwadkar, S., K. Pugh, A. Gupta and S. Ingole (2015). "Nanoparticles in the environment: occurrence, distribution, and risks". Journal of Hazardous Toxic and Radioactive Waste **19**(3): 1–8.

Li, N. J., C. J. Sun, J. J. Jiang, A. Q. Wang, C. Wang, Y. Shen, B. N. Huang, C. C. An, B. Cui, X. Zhao, C. X. Wang, F. Gao, S. S. Zhan, L. Guo, Z. H. Zeng, L. Zhang, H. X. Cui and Y. Wang (2021). "Advances in controlled-release pesticide formulations with improved efficacy and targetability". Journal of Agricultural and Food Chemistry **69**(43): 12579–12597.

Ma, C. X., J. C. White, J. Zhao, Q. Zhao and B. S. Xing (2018). "Uptake of engineered nanoparticles by food crops: Characterization, mechanisms, and implications". Annual Review of Food Science and Technology Vol 9. M. P. Doyle and T. R. Klaenhammer. Palo Alto, Annual Reviews **9**: 129–153.

Malandrakis, A. A., N. Kavroulakis and C. V. Chrysikopoulos (2020). "Use of silver nanoparticles to counter fungicide-resistance in Monilinia fructicola". Science of the Total Environment **747**: 1–13.

Marchetti, L. and S. C. Andres (2021). "Use of nanocellulose in meat products". Current Opinion in Food Science **38**: 96–101.

McClements, D. J. (2019). Future Foods: How Modern Science is Transforming the Way We Eat. Cham, Switzerland, Springer Scientific.

Mikula, K., G. Izydorczyk, D. Skrzypczak, M. Mironiuk, K. Moustakas, A. Witek-Krowiak and K. Chojnacka (2020). "Controlled release micronutrient fertilizers for precision agriculture – A review". Science of the Total Environment **712**: 126365, 1–17.

Monisha, M. and T. G. Dhanalakshmi (2015). "A review on precision agriculture and its farming methods". Research Journal of Pharmaceutical Biological and Chemical Sciences **6**(3): 1142–1153.

Monreal, C. M., M. DeRosa, S. C. Mallubhotla, P. S. Bindraban and C. Dimkpa (2016). "Nanotechnologies for increasing the crop use efficiency of fertilizer-micronutrients". Biology and Fertility of Soils **52**(3): 423–437.

Monteiro, A., S. Santos and P. Goncalves (2021). "Precision agriculture for crop and livestock farming-brief review". Animals **11**(8): 2345, 1–12. doi.org/10.3390/ani11082345

Montoya, Ú., R. Zuluaga, C. Castro, L. Vélez and P. Gañán (2019). "Starch and Starch/Bacterial Nanocellulose Films as Alternatives for the Management of Minimally Processed Mangoes". Starch-Stärke **71**(5–6): 1800120.

Mossa, A.-T. H., S. M. M. Mohafrash, E.-S. H. E. Ziedan, I. S. Abdelsalam and A. F. Sahab (2021). "Development of eco-friendly nanoemulsions of some natural oils and evaluating of its efficiency against postharvest fruit rot fungi of cucumber". Industrial Crops and Products **159**: 113049, 1–12. doi.org/10.1016/j.indcrop.2020.113049

Mossa, A. T. H., S. I. Afia, S. M. M. Mohafrash and B. A. Abou-Awad (2018). "Formulation and characterization of garlic (Allium sativum L.) essential oil nanoemulsion and its acaricidal activity on eriophyid olive mites (Acari: Eriophyidae)". Environmental Science and Pollution Research **25**(11): 10526–10537.

Mostafalou, S. and M. Abdollahi (2013). "Pesticides and human chronic diseases: Evidences, mechanisms, and perspectives". Toxicology and Applied Pharmacology **268**(2): 157–177.

Nasr-Eldahan, S., A. Nabil-Adam, M. A. Shreadah, A. M. Maher and T. E. A. Ali (2021). "A review article on nanotechnology in aquaculture sustainability as a novel tool in fish disease control". Aquaculture International **29**(4): 1459–1480.

Naylor, R. L., R. W. Hardy, A. H. Buschmann, S. R. Bush, L. Cao, D. H. Klinger, D. C. Little, J. Lubchenco, S. E. Shumway and M. Troell (2021). "A 20-year retrospective review of global aquaculture". Nature **591**(7851): 551-+.

Pavela, R. and G. Benelli (2016). "Essential oils as ecofriendly biopesticides? challenges and constraints". Trends in Plant Science **21**(12): 1000–1007.

Peixoto, S., I. Henriques and S. Loureiro (2021). "Long-term effects of Cu(OH)(2) nanopesticide exposure on soil microbial communities". Environmental Pollution **269**: 116113, 1–14. doi.org/10.1016/j.envpol.2020.116113

Peters, R. J. B., H. Bouwmeester, S. Gottardo, V. Amenta, M. Arena, P. Brandhoff, H. J. P. Marvin, A. Mech, F. B. Moniz, L. Q. Pesudo, H. Rauscher, R. Schoonjans, A. K. Undas, M. V. Vettori, S. Weigel and K. Aschberger (2016). "Nanomaterials for products and application in agriculture, feed and food". Trends in Food Science & Technology **54**: 155–164.

Peydayesh, M., S. Bolisetty, T. Mohammadi and R. Mezzenga (2019). "Assessing the binding performance of amyloid-carbon membranes toward heavy metal ions". Langmuir **35**(11): 4161–4170.

Poore, J. and T. Nemecek (2018). "Reducing food's environmental impacts through producers and consumers". Science **360**(6392): 987-+.

Pradhan, S. and D. R. Mailapalli (2020). "Nanopesticides for Pest Control". Sustainable Agriculture Reviews 40 E. Lichtfouse. **40**: 43–74.

Qi, W. H., J. J. Wu, Y. Shu, H. Q. Wang, W. L. Rao, H. N. Xu and Z. S. Zhang (2020). "Microstructure and physiochemical properties of meat sausages based on nanocellulose-stabilized emulsions". International Journal of Biological Macromolecules **152**: 567–575.

Raliya, R., J. C. Tarafdar and P. Biswas (2016). "Enhancing the mobilization of native phosphorus in the mung bean rhizosphere using zno nanoparticles synthesized by soil fungi". Journal of Agricultural and Food Chemistry **64**(16): 3111–3118.

Rao, J. J., B. C. Chen and D. J. McClements (2019). "Improving the efficacy of essential oils as antimicrobials in foods: mechanisms of action". Annual Review of Food Science and Technology Vol 10. M. P. Doyle and D. J. McClements **10**: 365–387.

Rodrigues, S. M., P. Demokritou, N. Dokoozlian, C. O. Hendren, B. Karn, M. S. Mauter, O. A. Sadik, M. Safarpour, J. M. Unrine, J. Viers, P. Welle, J. C. White, M. R. Wiesner and

G. V. Lowry (2017). "Nanotechnology for sustainable food production: promising opportunities and scientific challenges". Environmental Science-Nano **4**(4): 767–781.

Roy, S., P. K. Dikshit, K. C. Sherpa, A. Singh, S. Jacob and R. C. Rajak (2021). "Recent nanobiotechnological advancements in lignocellulosic biomass valorization: A review". Journal of Environmental Management **297**: 113422, 1–15. doi.org/10.1016/j.jenvman.2021.113422

Saleem, M. H., J. Potgieter and K. M. Arif (2021). "Automation in Agriculture by Machine and Deep Learning Techniques: A Review of Recent Developments". Precision Agriculture **22**(6): 2053–2091.

Sani, M. A., M. Azizi-Lalabadi, M. Tavassoli, K. Mohammadi and D. J. McClements (2021). "Recent Advances in the Development of Smart and Active Biodegradable Packaging Materials". Nanomaterials **11**(5): 1331, 1–19. doi: 10.3390/nano11051331.

Schappo, F. B., C. D. F. Ribeiro, M. Farina and I. L. Nunes (2022). "The toxicity of oil nanoparticles: a review focused on food science". Food Reviews International, 1–14. doi.org/10.1080/87559129.2021.2008954

Servin, A., W. Elmer, A. Mukherjee, R. De la Torre-roche, H. Hamdi, J. C. White, P. Bindraban and C. Dimkpa (2015). "A review of the use of engineered nanomaterials to suppress plant disease and enhance crop yield". Journal of Nanoparticle Research **17**(2): 92, 1–23. doi.org/10.1007/s11051-015-2907-7

Shafi, U., R. Mumtaz, J. Garcia-Nieto, S. A. Hassan, S. A. R. Zaidi and N. Iqbal (2019). "Precision agriculture techniques and practices: from considerations to applications". Sensors **19**(17): 3796, 1–14. doi.org/10.3390/s19173796

Shah, B. R. and J. Mraz (2020). "Advances in nanotechnology for sustainable aquaculture and fisheries". Reviews in Aquaculture **12**(2): 925–942.

Shang, Y. F., M. K. Hasan, G. J. Ahammed, M. Q. Li, H. Q. Yin and J. Zhou (2019). "Applications of nanotechnology in plant growth and crop protection: a review". Molecules **24**(14): 2558, 1–21. doi: 10.3390/molecules24142558.

Simonin, M., B. P. Colman, W. Y. Tang, J. D. Judy, S. M. Andersono, C. M. Bergemanno, J. D. Rocca, J. M. Unrine, N. Cassar and E. S. Bernhardt (2018). "Plant and microbial responses to repeated CU(OH)2 nanopesticide exposures under different fertilization levels in an agro-ecosystem". Frontiers in Microbiology **9**: 1769, 1–12. doi:10.3389/fmicb.2018.01769

Singh, A., N. Dhiman, A. K. Kar, D. Singh, M. P. Purohit, D. Ghosh and S. Patriaik (2020). "Advances in controlled release pesticide formulations: Prospects to safer integrated pest management and sustainable agriculture". Journal of Hazardous Materials **385**: 121525, 1–23. doi: 10.3390/molecules24142558

Singh, A., A. K. Kar, D. Singh, R. Verma, N. Shraogi, A. Zehra, K. Gautam, S. Anbumani, D. Ghosh and S. Patnaik (2022). "pH-responsive eco-friendly chitosan modified cenosphere/alginate composite hydrogel beads as carrier for controlled release of Imidacloprid towards sustainable pest control". Chemical Engineering Journal **427**: 131215, 1–13. doi/10.1016/j.cej.2021.131215

Singh, R. P., R. Handa and G. Manchanda (2021). "Nanoparticles in sustainable agriculture: An emerging opportunity". Journal of Controlled Release **329**: 1234–1248.

Soares, R. M. D., N. M. Siqueira, M. P. Prabhakaram and S. Ramakrishna (2018). "Electrospinning and electrospray of bio-based and natural polymers for biomaterials development". Materials Science & Engineering C-Materials for Biological Applications **92**: 969–982.

Srivastava, A. K., A. Dev and S. Karmakar (2017). "Nanosensors for food and agriculture". Nanoscience in Food and Agriculture 5 S. Ranjan, N. Dasgupta and E. Lichtfouse **26**: 41–79.

Stavrinidou, E., G. Dufil, I. Bernacka-Wojcik and A. Armada-Moreira (2022). "Plant bioelectronics and biohybrids: the growing contribution of organic electronic and carbon-based materials". Chemical Reviews **122**(4), 4847–4883.

Sun, H. Z., Y. L. Zou, H. Y. Kaw, L. Y. Wang, G. Wang, J. L. Zhou, L. Y. Meng and D. H. Li (2021). "Carbon nanofibers-based nanoconfined liquid phase filtration for the rapid removal of chlorinated pesticides from ginseng extracts". Journal of Agricultural and Food Chemistry **69** (32): 9434–9442.

Swift, T. A., D. Fagan, D. Benito-Alifonso, S. A. Hill, M. L. Yallop, T. A. A. Oliver, T. Lawson, M. C. Galan and H. M. Whitney (2021). "Photosynthesis and crop productivity are enhanced by glucose-functionalised carbon dots". The New phytologist **229**(2): 783–790.

Thakur, D., Y. Kumar, A. Kumar and P. K. Singh (2019). "Applicability of wireless sensor networks in precision agriculture: a review". Wireless Personal Communications **107**(1): 471–512.

Toksha, B., V. A. M. Sonawale, A. Vanarase, D. Bornare, S. Tonde, C. Hazra, D. Kundu, A. Satdive, S. Tayde and A. Chatterjee (2021). "Nanofertilizers: A review on synthesis and impact of their use on crop yield and environment". Environmental Technology & Innovation **24**: 101986, pages 1–15. doi/10.1016/j.eti.2021.101986

Usman, M., M. Farooq, A. Wakeel, A. Nawaz, S. A. Cheema, H. U. Rehman, I. Ashraf and M. Sanaullah (2020). "Nanotechnology in agriculture: Current status, challenges and future opportunities". Science of The Total Environment **721**: 137778, 1–19. doi.org/10.1016/j.scitotenv.2020.137778

Vanti, G. L., S. Masaphy, M. Kurjogi, S. Chakrasali and V. B. Nargund (2020). "Synthesis and application of chitosan-copper nanoparticles on damping off causing plant pathogenic fungi". International Journal of Biological Macromolecules **156**: 1387–1395.

Vanti, G. L., V. B. Nargund, K. N. Basavesha, R. Vanarchi, M. Kurjogi, S. I. Mulla, S. Tubaki and R. R. Patil (2019). "Synthesis of Gossypium hirsutum-derived silver nanoparticles and their antibacterial efficacy against plant pathogens". Applied Organometallic Chemistry **33**(1): e4630, 1–17. doi/10.1002/aoc.4630

Vejan, P., T. Khadiran, R. Abdullah and N. Ahmad (2021). "Controlled release fertilizer: A review on developments, applications and potential in agriculture". Journal of Controlled Release **339**: 321–334.

Velasquez-Cock, J., A. Serpa, L. Velez, P. Ganan, C. G. Hoyos, C. Castro, L. Duizer, H. D. Goff and R. Zuluaga (2019). "Influence of cellulose nanofibrils on the structural elements of ice cream". Food Hydrocolloids **87**: 204–213.

Veleirinho, B. and J. A. Lopes-da-Silva (2009). "Application of electrospun poly(ethylene terephthalate) nanofiber mat to apple juice clarification". Process Biochemistry **44**(3): 353–356.

Velez, Y. S. P., R. Carrillo-Gonzalez and M. D. A. Gonzalez-Chavez (2021). "Interaction of metal nanoparticles-plants-microorganisms in agriculture and soil remediation". Journal of Nanoparticle Research **23**(9): 206, 1–21. doi/10.1007/s11051-021-05269-3

Verma, K. K., X. P. Song, A. Joshi, D. D. Tian, V. D. Rajput, M. Singh, J. Arora, T. Minkina and Y. R. Li (2022). "Recent Trends in Nano-Fertilizers for Sustainable Agriculture under Climate Change for Global Food Security". Nanomaterials **12**(1): 173, 1–14. doi/10.3390/nano12010173

Vigneshwaran, N., D. M. Kadam and S. Patil (2019). Nanomaterials for Active and Smart Packaging of Food.

Vilarinho, F., A. S. Silva, M. F. Vaz and J. P. Farinha (2018). "Nanocellulose in green food packaging". Critical Reviews in Food Science and Nutrition **58**(9): 1526–1537.

Wang, P., E. Lombi, F. J. Zhao and P. M. Kopittke (2016). "Nanotechnology: A New Opportunity in Plant Sciences". Trends in Plant Science **21**(8): 699–712.

Wani, A. H., M. Amin, M. Shahnaz and M. A. Shah (2012). "Antimycotic Activity of Nanoparticles of MgO, FeO and ZnO on some Pathogenic Fungi". International Journal of Manufacturing Materials and Mechanical Engineering **2**(4): 59–70.

WEF (2019). Innovation with a Purpose: The role of technology innovation in accelerating food systems transformation. Geneva, Switzerland, World Economic Forum: 1–42.

Willett, W., J. Rockstrom, B. Loken, M. Springmann, T. Lang, S. Vermeulen, T. Garnett, D. Tilman, F. DeClerck, A. Wood, M. Jonell, M. Clark, L. J. Gordon, J. Fanzo, C. Hawkes, R. Zurayk, J. A. Rivera, W. De Vries, L. M. Sibanda, A. Afshin, A. Chaudhary, M. Herrero, R. Agustina, F. Branca, A. Lartey, S. G. Fan, B. Crona, E. Fox, V. Bignet, M. Troell, T. Lindahl, S. Singh, S. E. Cornell, K. S. Reddy, S. Narain, S. Nishtar and C. J. L. Murray (2019). "Food in the Anthropocene: the EAT-Lancet Commission on healthy diets from sustainable food systems". Lancet **393**(10170): 447–492.

Wink, M. (2022). "Current Understanding of Modes of Action of Multicomponent Bioactive Phytochemicals: Potential for Nutraceuticals and Antimicrobials". Annual Review of Food Science and Technology **13**(1): 337–359.

Wong, M. H., J. P. Giraldo, S. Y. Kwak, V. B. Koman, R. Sinclair, T. T. S. Lew, G. Bisker, P. W. Liu and M. S. Strano (2017). "Nitroaromatic detection and infrared communication from wild-type plants using plant nanobionics". Nature Materials **16**(2): 264–272.

Xiang, W., X. Y. Zhang, J. J. Chen, W. X. Zou, F. He, X. Hu, D. C. W. Tsang, Y. S. Ok and B. Gao (2020). "Biochar technology in wastewater treatment: A critical review". Chemosphere **252**: 126539, 1–19. doi/10.1016/j.chemosphere.2020.126539

Xiang, Y. B., G. L. Zhang, C. W. Chen, B. Liu, D. Q. Cai and Z. Y. Wu (2018). "Fabrication of a pH-Responsively Controlled-Release Pesticide Using an Attapulgite-Based Hydrogel". Acs Sustainable Chemistry & Engineering **6**(1): 1192–1201.

Xu, Z. L., T. Tang, Q. Lin, J. Z. Yu, C. P. Zhang, X. P. Zhao, M. Kah and L. X. Li (2022). "Environmental risks and the potential benefits of nanopesticides: a review". Environmental Chemistry Letters **20**(3), 2097–2108.

Yao, M. F., J. J. Xie, H. J. Du, D. J. McClements, H. Xiao and L. J. Li (2020). "Progress in microencapsulation of probiotics: A review". Comprehensive Reviews in Food Science and Food Safety **19**(2): 857–874.

Zaynab, M., M. Fatima, S. Abbas, Y. Sharif, M. Umair, M. H. Zafar and K. Bahadar (2018). "Role of secondary metabolites in plant defense against pathogens". Microbial Pathogenesis **124**: 198–202.

Zhang, Z. Y., Y. B. Tan and D. J. McClements (2021). "Investigate the adverse effects of foliarly applied antimicrobial nanoemulsion (carvacrol) on spinach". Lwt-Food Science and Technology **141**: 110936, 1–9. doi:10.1016/j.lwt.2021.110936

Zhao, L., Y. Huang, C. Hannah-Bick, A. N. Fulton and A. A. Keller (2016). "Application of metabolomics to assess the impact of Cu(OH)(2) nanopesticide on the nutritional value of lettuce (Lactuca sativa): Enhanced Cu intake and reduced antioxidants". Nanoimpact **3–4**: 58–66.

Zhou, H. L., J. N. Liu, T. T. Dai, J. L. M. Mundo, Y. B. Tan, L. Bai and D. J. McClements (2021). "The gastrointestinal fate of inorganic and organic nanoparticles in vitamin D-fortified plant-based milks". Food Hydrocolloids **112**: 106310, 1–8. doi/10.1016/j.foodhyd.2020.106310

Zhou, H. L., Y. B. Tan, S. S. Lv, J. N. Liu, J. L. M. Mundo, L. Bai, O. J. Rojas and D. J. McClements (2020). "Nanochitin-stabilized pickering emulsions: Influence of nanochitin on lipid digestibility and vitamin bioaccessibility". Food Hydrocolloids **106**: 05878, 1–10. doi:10.1016/j.foodhyd.2020.105878

Chapter 4
Applications in food and nutrition

4.1 Introduction

The use of nanotechnology to improve the safety, quality, healthiness, and shelf-life of foods has been the focus of numerous food researchers since the early 2000s (McClements 2019). Most of these applications involve the use of food-grade nanoparticles, which may be either organic or inorganic depending on the required functional attributes (Chapter 1). As well as academic research in this area, many food companies have explored the potential of nanotechnology for enhancing the functional performance of their food and beverage products. In this section, several of the most promising applications of nanotechnology for improving food and nutrition are highlighted.

4.2 Improving food quality

In this section, an overview of the various ways nanoparticles can be used to modify the quality attributes of foods is given, such as their optical, rheological, and stability characteristics. Foods with new or improved quality attributes can often be created by utilizing the unique structural and physicochemical properties of small particles.

4.2.1 Appearance

The scattering of light waves by nanoparticles depends strongly on their dimensions relative to the wavelength of light (Figure 2.4). As discussed in Chapter 2, a suspension of nanoparticles goes from optically clear to optically opaque as the diameter of the nanoparticles is increased from much smaller than the wavelength of light to comparable to the wavelength of light, e.g., by increasing the diameter from around 10 to 1000 nm. Consequently, the dimensions of nanoparticles can be controlled to modulate the optical properties of foods to make them more desirable to consumers. In this section, we therefore consider the impact of nanoparticles on the optical properties of foods and beverages.

The optical properties of nanoparticle suspensions, such as their color and opacity, are governed by the selective absorption and scattering of light waves by the chromophores and particles they contain, respectively (McClements 2002b, McClements 2002a, McClements 2015b). The color (e.g., redness, blueness, and yellowness) of a nanoparticle suspension (or a food containing it) is a result of the absorption of

https://doi.org/10.1515/9783110788457-004

light waves over specific wavelength ranges caused by functional groups on the chromophores that selectively absorb visible light. For instance, carotenoids contain many conjugated double bonds that selectively absorb light waves from the green to violet regions of the electromagnetic spectrum, thereby leading to bright red, orange, or yellow colors. The opacity, cloudiness, or turbidity of a nanoparticle suspension (or a food containing it) depends on the strength of light scattering by the particles, which is governed by their dimensions relative to the wavelength of light (Figure 2.4). When the particles have diameters below about 50 nm, the light scattering is very weak, and the suspensions appear optically clear. As the particle diameter increases from around 50 to 200 nm, the intensity of the light scattering increases and the suspension appears increasingly turbid. When the particle size is increased further, the intensity of the light scattering decreases, which causes the turbidity to decrease. An understanding of the factors impacting the optical properties of nanoparticle suspensions is important when hydrophobic substances, such as colors, flavors, preservatives, vitamins, or nutraceuticals, need to be incorporated into aqueous-based food or beverage products. Typically, these hydrophobic substances are first incorporated into small oil droplets that are then incorporated into the products. For foods or beverages where a clear product is required (like soft drinks, fortified waters, or gummies), it is important to use nanoparticles that are relatively small (<50 nm). Conversely, for products that should be cloudy or opaque (such as milks or milk analogs), it is important to use nanoparticles with intermediate dimensions (200 to 500 nm).

Experimentally, the appearances of nanoparticle suspensions or the foods containing them are determined by measuring the transmittance or reflectance of visible light as a function of wavelength. Transmittance measurements are more suitable for analyzing clear or slightly turbid systems, whereas reflectance measurements are more suitable for turbid or opaque systems (Hutchings 1999). These measurements are then often converted into color coordinates by the instrument, e.g., L^*, a^*, and b^* values (McClements 2002b, McClements 2002a). Here, L^* is known as the lightness, which varies from 0 (pure black) to 100 (pure white). The a^* and b^* values provide information about the chromaticity of the material: $+a^*$ is redness, $-a^*$ is greenness, $+b^*$ is yellowness, and $-b^*$ is blueness. The opacity can then be specified in terms of the lightness (L^*), while the overall color intensity can be specified in terms of the chroma: $C = (a^{*2} + b^{*2})^{1/2}$. Mathematical models based on light scattering theory have been developed to link the characteristics of nanoparticles (concentration, diameter, and refractive index) and chromophore characteristics (concentration and absorption spectrum) to the L^*, a^*, and b^* values (McClements 2002b, McClements 2002a). The appearance of nanoparticle suspensions and foods can be characterized by taking a digital photograph, using a dedicated colorimeter, or using a UV-visible spectrophotometer with transmission and/or reflectance options (Hutchings 1999).

Various kinds of food grade nanoparticles have been used to control the optical properties of foods. Titanium dioxide nanoparticles are widely used as brightening or lightening agents in several kinds of foods and beverages because they have

high refractive indices and diameters similar to the wavelength of light and therefore scatter light strongly (McClements et al. 2017). As discussed earlier, oil-soluble
active agents like vitamins, nutraceuticals, preservatives, colors, or flavors, can be
incorporated into nanosized oil droplets so they can be introduced into foods and
beverages (Akhavan et al. 2018, McClements 2020c).

4.2.2 Texture

The small dimensions of nanoparticles can also be important for modifying the textural attributes of certain kinds of foods (Ikeda and Zhong 2012). In general, the introduction of nanoparticles into a food product changes its textural attributes, with the
nature of the change depending on the concentration, size, shape, and interactions
of the particles (McClements 2015b). Initially, the viscosity of a nanoparticle suspension increases gradually with increasing nanoparticle concentration but then it increases much more steeply. Eventually, the nanoparticles begin to interact strongly
with each other, which leads to solid-like characteristics. The small size of nanoparticles usually has a more pronounced impact on the rheological properties of materials at relatively high concentrations, where particle-particle interactions become
more important. For instance, at the same overall particle concentration, small particles often lead to a much higher viscosity than larger ones in systems where there is
either a strong attraction or repulsion between the particles (Figure 4.1).

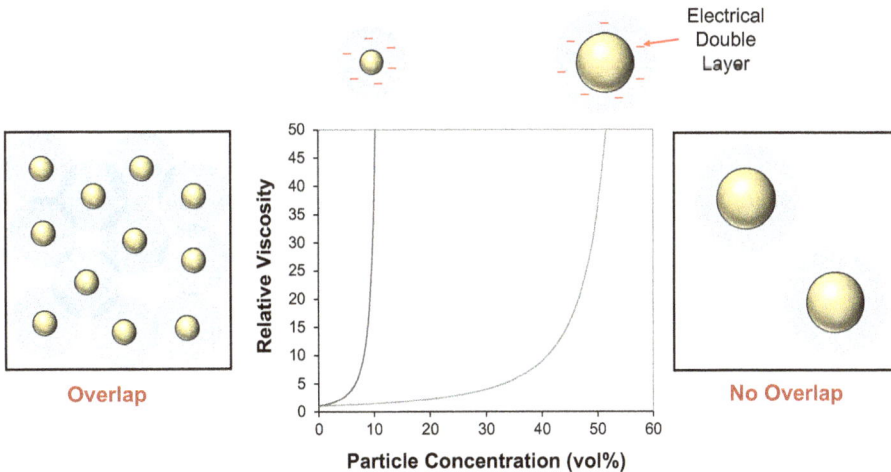

Figure 4.1: Decreasing the size of electrically charged particles can lead to a higher viscosity
at low particle concentrations due to overlap of the electrical double layers.

Mathematical models have been developed to relate the rheological properties of nanoparticle suspensions to their characteristics (McClements 2015c). To a first approximation, the Einstein equation can be used to predict the shear viscosity of a dilute suspension of spherical nanoparticles suspended in an ideal fluid:

$$\eta = \eta_1(1 + 2.5\phi) \tag{4.1}$$

In this equation, η_1 is the shear viscosity of the fluid surrounding the nanoparticles and ϕ is the dispersed phase volume fraction of the nanoparticles. This equation predicts that the viscosity of dilute nanoparticle suspensions should rise linearly with increasing nanoparticle concentration (Figure 4.1). Interestingly, this equation suggests that particle size does not have a major impact on the shear viscosity of dilute systems. The Einstein equation gives reliable predictions of the viscosity of nanoparticle suspensions up to particle concentrations of about 5% for non-interacting spheres (Larson 1999).

The shear viscosity of more concentrated nanoparticle suspensions can be described by a semi-empirical equation (McClements 2015c):

$$\eta = \eta_1\left(1 - \frac{\phi_{eff}}{\phi_C}\right)^{-2} \tag{4.2}$$

Here, ϕ_{eff} is the effective volume fraction of the nanoparticles and ϕ_C is the critical volume fraction of the nanoparticles where they become close-packed, which has a value around 0.65. As the nanoparticle concentration is raised toward this critical volume fraction, the nanoparticles become so closely packed together that the suspension gains solid-like characteristics, which leads to a large increase in viscosity and then conversion to a gel (Figure 4.1).

For nanoparticles, the effective volume fraction (ϕ_{eff}) can be much greater than their actual volume fraction (ϕ), especially when there are strong attractive or repulsive interactions between them. For instance, when there are steric or electrostatic repulsive forces between the nanoparticles, the effective volume fraction is given by:

$$\phi_{eff} = \phi\left(1 + \frac{2\delta}{d}\right)^3 \tag{4.3}$$

Here, δ is the effective thickness of the interfacial region around the nanoparticles. For steric repulsion, this value is equal to the thickness of the interfacial layer formed by the substances adsorbed to the surfaces of the nanoparticles. For electrostatic repulsion, this value is related to the Debye screening length, which is the distance that the electrostatic potential extends into the surrounding fluids. Typically, the Debye screening length decreases with increasing ionic strength, ranging from around 30 nm in a 0.1 mM NaCl solution to 0.3 nm in a 1 M NaCl solution (McClements 2015b). When nanoparticles have a high surface potential (strong negative or positive charge) and they are suspended in a fluid with a low ionic strength, the range of the electrostatic repulsion between them is relatively long. This means that the nanoparticles cannot approach very close to each other, which increases their effective volume fraction. As a result, the viscosity can increase more rapidly with increasing particle concentration for small particles (Figure 4.1). This phenomenon may be useful for creating highly viscous or gelled food products at relatively low nanoparticle concentrations.

There may also be a large increase in viscosity due to the small dimensions of nanoparticles when there are strong attractive forces between them (McClements 2015b). For instance, research has shown that the viscosity of suspensions containing a mixture of positively charged and negatively charged particles increases as their diameters decrease (Mao and McClements 2012). This effect is because the oppositely charged particles are strongly attracted to each other and can form a 3D network, thereby giving solid-like properties. Smaller particles can form a 3D network that extends throughout the entire system at lower concentrations than larger particles. The impact of the rheological properties of nanoparticle suspensions on their particle size and interactions means that these parameters can be controlled to create advanced materials with novel or improved textural attributes. For instance, it may be possible to create reduced calorie products that are highly viscous or semi-solid (like sauces or dressings) using lower fat contents by using nanosized fat droplets rather than conventional-sized ones.

Several kinds of instruments are typically used to characterize the rheological properties of nanoparticle suspensions (McClements 2015b). The type of instrument used for a particular application depends on the state of the material being analyzed (e.g., liquid, solid, or viscoelastic), the type of force applied to the material (e.g., compression, shear, and/or extensional), and the rheological parameter being determined in the experiment (e.g., viscosity, elastic modulus, yield stress, fracture stress, and/or fracture strain) (McClements 2015a). Typically, the sophistication and versatility of the instrument depends on its cost. For quality assurance applications, the instruments are often cheap, simple, and rapid. Conversely, for research and development applications, the instruments are often much more expensive, complex, and time-consuming but they provide more fundamental information about the rheology of the material being analyzed. Moreover, the data obtained can more easily be compared with that of other researchers. The most common rheological instruments used to test nanoparticle suspensions are shear and compression rheometers (Figures 4.2 and 4.3). Shear rheometers are most suitable for fluid or semi-solid samples. They usually involve applying a known shear stress parallel to the surface of a sample and then measuring the rate of strain. For example, a fluid sample may be placed into the cup of a concentric cylinder-based measurement cell and then the bob is lowered into it (Figure 4.2). A shear stress is then applied to the bob and the rate it rotates is measured by the instrument: the faster it rotates, the lower the viscosity. The shear stress *versus* shear rate profile is measured and reported

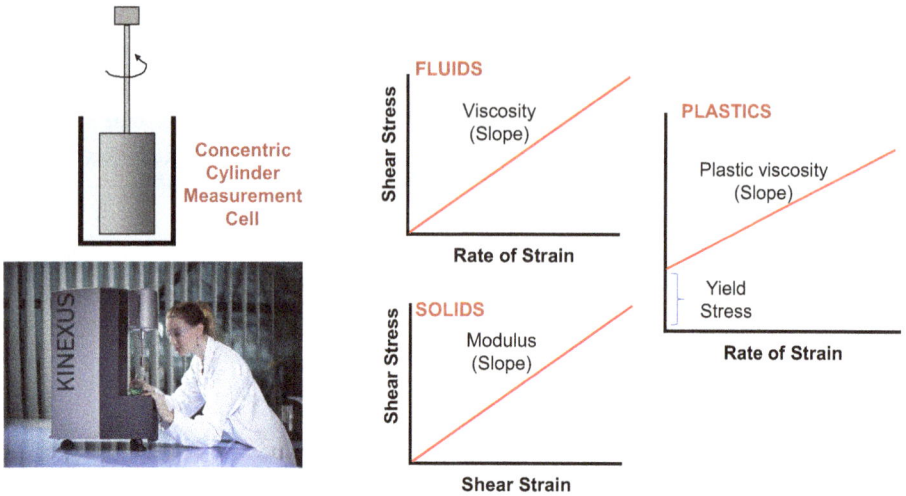

Figure 4.2: The apparent shear viscosity or elastic modulus of fluid or solid nanomaterials can be characterized by dynamic shear rheometers. Image of dynamic shear rheometer kindly supplied by Philip Rolfe (Netzsch).

by the instrument. The apparent shear viscosity can then be determined as the slope of this curve, which may change with shear rate (non-ideal behavior). For plastic materials, the stress *versus* rate of strain curve is still determined, but in this case, flow is only observed when the stress exceeds a critical value (known as the yield stress). For viscoelastic materials, a sinusoidal stress is applied, the resulting sinusoidal strain is measured, and then the complex shear modulus and phase angle are calculated.

Compression rheometers are more suitable for testing the properties of semi-solid or solid materials (Figure 4.3). They typically involve applying a known compressional stress perpendicular to the surface of a sample and measuring the strain. For example, a solid sample of known dimensions is placed on a platform and then a probe is lowered downwards. The force acting on the probe is then measured as a function of distance and a stress (force per unit surface area) *versus* stain (deformation divided by original length) of the sample is measured. The stress *versus* strain profile of the sample can then be used to obtain information such as the elastic modulus, fracture stress, and fracture strain. In some cases, the test material is compressed/decompressed twice and then the force *versus* time profile is acquired. Parameters such as hardness, springiness, chewiness, and cohesiveness can then be calculated from these profiles, which is known as texture profile analysis (TPA). This type of test is often carried out to provide some insights into how a food may behave inside the human mouth.

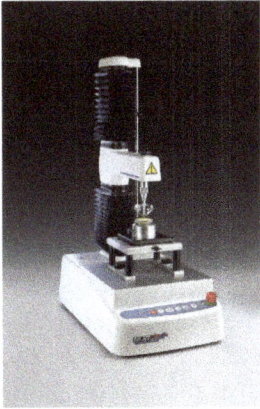

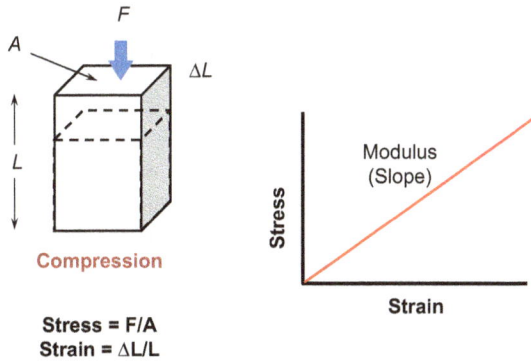

Figure 4.3: The textural attributes of solid or semi-solid nanomaterials can be characterized using texture profile analysis with a compression instrument. Photograph of texture analyzer kindly supplied by Marc Johnson, of Texture Technologies Corp. and used with permission.

4.2.3 Flavor

In some applications, nanoparticles may be useful for controlling the retention and release of flavor molecules, which means they can be used to alter the dynamic flavor profile of foods (McClements 2015c). Oil-soluble flavors that cannot normally be dispersed in water can be encapsulated inside nanosized oil droplets so they can then easily be introduced into aqueous-based food and beverage products. As mentioned earlier, the size of the oil droplets may be selected based on the required appearance of the final product: small droplets ($d < 50$ nm) for a clear product or larger droplets ($d = 200–500$ nm) for a cloudy or opaque product. The dimensions of nanoparticles are so small that they are not suitable for controlling the release of encapsulated flavor molecules (McClements 2015b). Typically, flavor molecules are released on the order of a second or much less because of the very short distance they must diffuse out of the small particles (like oil droplets). However, a controlled flavor release can be obtained by encapsulating flavor-loaded oil droplets within larger hydrogel beads. Typically, the larger the dimensions of the hydrogel beads used, the slower the release of the encapsulated flavors is. This approach has been utilized to control the release of hydrophobic garlic flavors during the cooking of foods (Wang et al. 2019).

4.2.4 Shelf life

Nanotechnology can also be utilized to extend the shelf life of some foods, thereby improving food quality, and reducing food waste. Foods are prone to contamination by spoilage and pathogenic microorganisms that can increase food waste, as well as increasing the risk of foodborne illness and death. The adverse effects of microbial contamination can be reduced by proper processing, packaging, and handling of foods, as well as by using various kinds of preservatives, such as antimicrobials and antioxidants. The utilization and efficacy of these preservatives can often be improved by encapsulating them within nanoparticles. For instance, oil-soluble antioxidants or antimicrobial agents can be incorporated into nanosized oil droplets, which can then be introduced into food or beverage products (McClements 2018b). The small size of these droplets means they are more resistant to gravitational separation and aggregation than conventional-sized droplets. Consequently, they tend to be more uniformly dispersed throughout the system. As a result, the preservatives can more easily access their site of action, such as the surfaces of the contaminating microbes. Converting food ingredients into nanosized particles, rather than conventional micro-sized ones, can also improve the physical stability of foods, such as their resistance to aggregation or gravitational separation (McClements 2015c).

Some kinds of nanoparticles used in foods are thermodynamically stable, such as micelles and microemulsions, which are typically formed from synthetic surfactants. Consequently, they should remain intact during storage if the environmental conditions are not altered (such as dilution or temperature), and no chemical or microbiological changes in the ingredients occurs. However, many kinds of nanoparticles are thermodynamically unstable, such as nanoemulsions, solid lipid nanoparticles, nanoliposomes, and biopolymer nanoparticles (McClements 2015c). As a result, they tend to breakdown over time due to various physicochemical processes, including gravitational separation (creaming/sedimentation), aggregation (flocculation/coalescence), or Ostwald ripening (Figure 4.4). Consequently, it is important to understand the origin of these different instability mechanisms and develop strategies to inhibit them. In this section, we therefore briefly discuss some of the most important ways that nanoparticle suspensions may breakdown and methods to prevent these undesirable changes from occurring.

4.2.4.1 Gravitational separation

The upward or downward movement of particles in foods due to gravitational forces is a common cause of instability in foods, leading to a reduction in food quality (McClements 2015b). Consequently, food manufacturers have developed a variety of strategies to reduce this problem. Gravitational separation may manifest itself as the upward movement of particles ("creaming") or the downward movement of particles ("sedimentation") depending on whether they are lighter or heavier than the surrounding fluids. Liquid oils are often less dense than water and so tend to cream, whereas solid fat-, protein-, and polysaccharide-based particles are often denser than water and so tend to sediment.

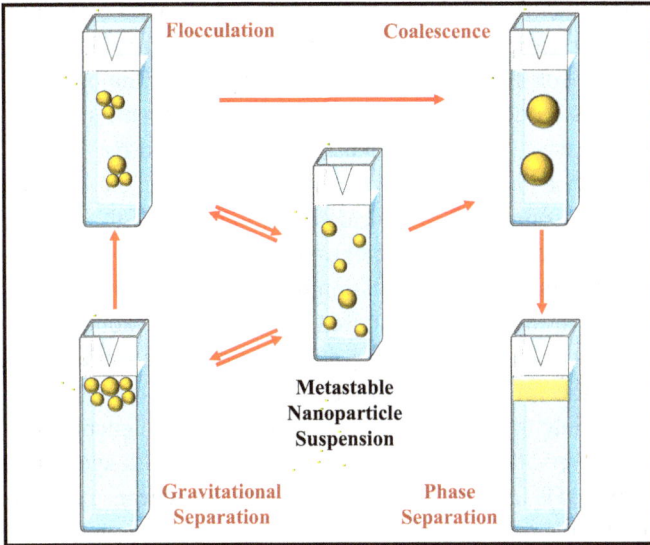

Figure 4.4: Schematic diagram of common instability mechanisms that occur in nanoparticle suspensions: gravitational separation, flocculation, coalescence, and phase separation.

The rate at which particles move due to gravity can be predicted using Stokes' law:

$$v = -\frac{gd^2(\rho_2 - \rho_1)}{18\eta_1} \tag{4.4}$$

In this equation, v is the velocity (m s^{-1}) that the particles move (which is positive for creaming and negative for sedimentation), g is the acceleration due to gravity (around 9.8 m s^{-2}), d is the particle diameter (m), ρ is the density (kg m^{-3}), η is the shear viscosity (kg m^{-1} s^{-1}), and the subscripts 1 and 2 refer to the fluid surrounding the particles and the particles themselves, respectively. The Stokes' equation is useful for highlighting the main factors impacting the resistance of particles to gravitational separation. This equation predicts that the velocity of particle movement decreases with decreasing particle diameter, decreasing density contrast, and increasing viscosity of the surrounding fluid. Stokes' law is derived assuming that the particles are rigid and spherical, that they do not interact with each other, and that Brownian motion is not important. Nevertheless, equations can be developed that include these different factors (McClements 2015a). Notably, Brownian motion is particularly important for relatively small particles ($d < 100$ nm), which is the case for many food-grade nanoparticles. Brownian motion tends to oppose creaming or sedimentation of nanoparticles because it has a disorganizing influence on their spatial distribution.

The rate of gravitational separation of nanoparticles can be measured using a variety of methods, including by simple visual observation, or by using analytical instruments specifically designed for this purpose (McClements 2015c). Commonly, instruments based on the backscattering or transmission of a laser beam as a function of sample height are used for this purpose (Figure 4.5). Typically, the backscattering increases and transmission decreases as the nanoparticle concentration at a specific location (height) increases. Other instruments can also be used for this purpose, such as magnetic resonance imaging, ultrasonic imaging, or X-ray tomography.

Nanoparticles are used in some commercial foods and beverages to increase their resistance to gravitational separation. The extremely small size of nanoparticles leads to a decrease in creaming or sedimentation because the velocity of particle movement is proportional to the diameter squared (Equation 4.4). Moreover, as the nanoparticle size decreases, the disorganizing effects of Brownian motion increase, which also reduces the rate of gravitational separation. As an example, nanosized oil droplets are widely used in soft drinks and beverages to encapsulate oil-soluble flavors, colors, preservatives, and vitamins in a form that will keep them stable. The use of nanoparticles is important for these applications because the viscosity of the surrounding fluid is relatively low, and so it is important to use small particles to inhibit gravitational separation based on Stokes' law (Piorkowski and McClements 2014).

4.2.4.2 Particle aggregation

Nanoparticles can be attracted or repelled from each other due to various kinds of colloidal interactions that operate between them, such as van der Waals, electrostatic, steric, hydrophobic, hydrogen bonding, depletion, and bridging interactions (McClements 2015b). When two or more nanoparticles come into contact with each other, for instance due to Brownian motion, stirring, or gravitational separation, they may either stick together or bounce off each other depending on whether the attractive or repulsive forces dominate (Israelachvili 2011). The sign, magnitude, and range of the colloidal interactions depend on the properties of the nanoparticles (such as their size, charge, and hydrophobicity), as well as the properties of the surrounding fluid (such as the ionic strength, pH, and additive type/concentration). As an example, the electrostatic repulsion between two nanoparticles tends to become stronger as their surface charge increases or the ionic strength decreases (McClements 2015b). The steric repulsion tends to increase as the interfacial coatings around the nanoparticles become thicker. The hydrophobic attraction tends to increase as the number of exposed non-polar groups on the surfaces of the nanoparticles increases. The depletion attraction tends to increase as the concentration of non-adsorbed polymer molecules in the surrounding fluid increases. Consequently, it is important to understand the nature of the colloidal interactions operating in a particular nanoparticle system to control their aggregation stability. Nanoparticles may undergo different kinds of aggregation when they come into contact with their neighbors, including flocculation, coalescence, and partial coalescence.

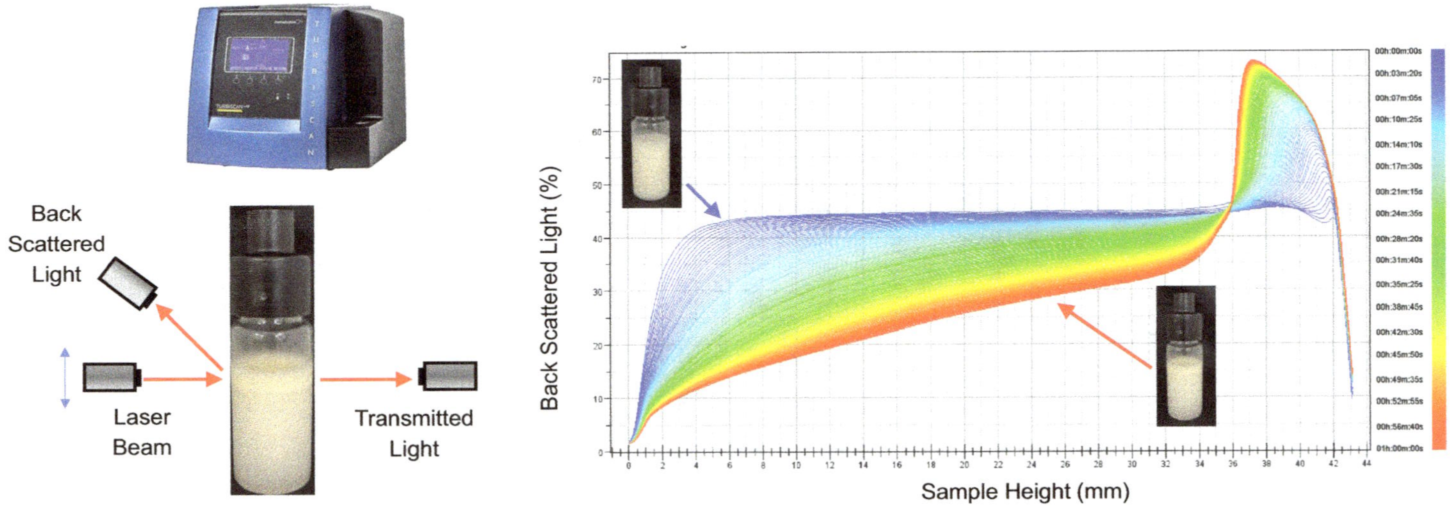

Figure 4.5: The stability of fluid plant-based foods to sedimentation or creaming can be conveniently monitored using instruments that measure the change in transmitted and backscattered light with sample height over time. Data kindly supplied by Dr. Hualu Zhou (UMASS). Image of Turbiscan instrument kindly provided by Formulaction.

Flocculation: This type of aggregation involves the formation of aggregates ("flocs") that contain two or more nanoparticles, with each individual nanoparticle retaining its original properties (Figure 4.4). Flocculation can be further categorized as weak or strong depending on the strength of the forces holding the nanoparticles together relatively to the thermal energy of the system (kT). Weak flocs can often be dissociated by simple stirring, whereas strong flocs cannot. For many applications, the flocculation of nanoparticles is undesirable since it increases the effective size of the particles in the system, which increases their gravitational separation rate (creaming or sedimentation). Moreover, flocculation can lead to an appreciable increase in the viscosity of a nanoparticle suspension, which may be either desirable or undesirable depending on the application (Section 4.2.2).

Coalescence: This kind of aggregation is only important for fluid nanoparticles (nanodroplets) that can merge after they contact each other (Figure 4.4). In this case, two or more nanodroplets encounter each other, merge together, and form a single larger droplet, which means that the interfacial layers around the nanodroplets must be disrupted. The coalescence of nanodroplets leads to an increase in gravitational separation (creaming or sedimentation) because of the increase in particle size, as predicted by Stokes' Law. Eventually, extensive coalescence of the oil droplets in oil-in-water nanoemulsions can lead to the creation of an oil layer at the top, which is usually known as phase separation or *oiling off*. In contrast, extensive coalescence of the water droplets in water-in-oil nanoemulsions can lead to the creation of a water layer at the bottom. The resistance of fluid nanoparticles to coalescence is mainly governed by the nature of the forces acting on them (e.g., colloidal, mechanical, or gravitational forces) and the robustness of their interfacial layers. Coalescence can be inhibited by preventing the nanoparticles from coming close together or by ensuring the interfacial layers are resistant to disruption.

Partial coalescence: This kind of aggregation occurs when the nanoparticles are only partially crystalline, i.e., they have some crystalline domains and some fluid domains. When two or more of these partially crystalline nanoparticles contact each other, they may form large clumps that are held together because a crystal in one particle can penetrate the fluid region of another particle (Fredrick et al. 2010, Petrut et al. 2016). Partial coalescence is typically important in suspensions of lipid nanoparticles that undergo a liquid-solid transition over the temperature range of usage. The tendency for partial coalescence to occur depends on the fraction of the lipid phase that is solidified, as well as the thickness and robustness of the interfacial layer around the lipid phase. If the lipid core is fully liquid or fully solid, then partial coalescence does not occur. A thick robust interfacial layer can inhibit the ability of the fat crystals to penetrate another nanoparticle.

Controlling nanoparticle aggregation: In many cases, nanoparticle aggregation is undesirable and so strategies should be used to prevent it. This can be achieved by increasing the repulsive interactions between the nanoparticles, or by reducing the attractive interactions between them. Steric repulsion can be strengthened by increasing the thickness of the interfacial layer. Electrostatic repulsion can be strengthened by altering the pH to increase the surface charge or by reducing the ionic strength. Hydrophobic attraction can be reduced by decreasing the number of exposed non-polar groups on the nanoparticle surfaces. Bridging and depletion attraction can be reduced by controlling the type and concentration of polymers in the system. In other cases, it is desirable to promote droplet aggregation in a controlled manner, which means that it is important to understand and control the different kinds of attractive and repulsive interactions in the system. The nature of the aggregation in a nanoparticle suspension can often be predicted using mathematical or computational models (McClements 2015b). These models are particularly useful for predicting the influence of specific nanoparticle properties (size, shape, concentration, charge, surface hydrophobicity, and interfacial thickness) and solution properties (pH, ionic strength, and additives) on the aggregation stability of nanoparticle suspensions.

4.2.4.3 Ostwald ripening

Another instability mechanism that is important in some nanoparticle suspensions is Ostwald ripening (Taylor 1998, Kabalnov 2001, Koroleva and Yurtov 2021). This process is important when the nanoparticles have a finite solubility in the surrounding fluid, which is important for flavor or essential oils, such as those used to make beverages or antimicrobial nanoemulsions. Ostwald ripening manifests itself with the growth of the larger particles at the expense of the smaller ones, which is a result of some of the core material diffusing between the intervening fluids (Figure 4.6).

To a first approximation, after steady state conditions have been reached, the Ostwald ripening rate is given by the following equation (Kabalnov and Shchukin 1992):

$$\frac{d(d_n)^3}{dt} = \frac{64\gamma V_m SD}{9RT} \tag{4.5}$$

Here, d_n is the number-weighted mean particle diameter, t is the time, γ is the interfacial tension, V_m is the molar volume of the core material, S is the equilibrium solubility of the core material in the surrounding fluid, D is the translation diffusion coefficient of the core material through the fluid surrounding the particles, R is the gas constant, and T is the absolute temperature. The most important parameter in this equation is the solubility of the core material in the surrounding fluid (S). As this value increases, the rate of Ostwald ripening increases. It is for this reason that nanoparticle growth due to Ostwald ripening is important in oil-in-water nanoemulsions fabricated from essential or flavor oils – these oils are relatively polar, which leads to a relatively high water solubility (Rao et al. 2019).

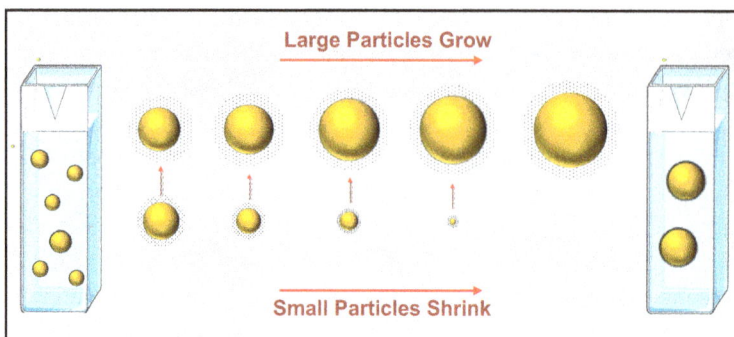

Figure 4.6: Schematic diagram of change in droplet size due to Ostwald ripening in nanoparticle suspensions, like nanoemulsions.

An effective strategy to inhibit Ostwald ripening in these kinds of nanoemulsions is to include substances known as ripening inhibitors in the oil phase. Ripening inhibitors are typically strongly hydrophobic molecules that have very low solubility in water, such as long-chain triacylglycerols like corn, sunflower, or olive oils (Wooster et al. 2008, Li et al. 2009). These substances demonstrate their stabilizing effect through an entropy of mixing phenomenon known as compositional ripening. Consider a polydisperse nanoemulsion that contains oil droplets containing polar oil (essential oil) and non-polar oil (ripening inhibitor). Immediately after preparation, all the droplets in the nanoemulsion have similar compositions but different sizes. Some of the polar oil molecules move from the smaller to the larger droplets because of Ostwald ripening. In contrast, the ripening inhibitor cannot diffuse between the oil droplets due to its very low water solubility. Consequently, the smaller droplets contain a reduced polar oil concentration, whereas the larger droplets contain an elevated polar oil concentration. This leads to a concentration gradient between the smaller and larger droplets in the system that is thermodynamically unfavorable because of entropy of mixing effects. As a result, there is a driving force for the polar molecules to move from the larger droplets (high concentration) to the smaller ones (low concentration), which counterbalances the Ostwald ripening effect. Typically, the ripening inhibitor concentration in the oil phase must exceed a critical value to completely stop Ostwald ripening, which depends on the molar mass and water solubility of the polar and non-polar oils used, and can be estimated using a suitable mathematical theory (Kabalnov and Shchukin 1992, Wooster et al. 2008).

4.3 Improving food nutrition

4.3.1 Modulation of macronutrient digestion

One of the reasons that has been postulated for the rising levels of diabetes and obesity in many countries over the past several decades has been overconsumption of ultra-processed foods (Lane et al. 2021). In the past, our ancient ancestors lived in environments where food was often scarce and the food that was available was often raw. As a result, the human gastrointestinal tract (GIT) evolved to efficiently breakdown tough fibrous foods and extract the nutrients. However, many of the foods consumed nowadays are highly processed so that their original cellular structures are disrupted, and the tough fibrous material has been removed. These foods include products such as white bread, cakes, cookies, soft drinks, sugary breakfast cereals, and snacks. These kinds of highly processed foods tend to be rapidly digested within the GIT, leading to spikes in the levels of glucose and lipids in the bloodstream. Moreover, they can change our appetite and satiety responses, such as our feelings of hunger and fullness. Highly processed foods can therefore cause dysregulation of the human metabolic and hormonal systems, which can promote overeating and chronic diseases like obesity, diabetes, and hypertension (McClements 2019, Lane et al. 2021). There is therefore interest in developing strategies to retard macronutrient digestion within the human gut. Several researchers have examined the possibility of using nanotechnology approaches to inhibit macronutrient digestion, with the aim of preventing or reducing these problems.

The impact of food-grade nanofibers on macronutrient digestion have been investigated recently, such as nanocellulose and nanochitin (Zhou et al. 2020, Liu and Kong 2021). Nanocellulose and nanochitin consist of thin fibrous materials isolated from cellulose-rich or chitin-rich sources, respectively. Nanocellulose can be isolated from wood and cotton, whereas nanochitin can be isolated from crab shells and some other sources. Several studies have shown that including nanofibers into foods can retard the digestion of macronutrients, e.g., nanocellulose has been shown to inhibit both lipid and starch digestion (DeLoid et al. 2018a, DeLoid et al. 2018b, Liu and Kong 2019, Guo et al. 2021, Liu and Kong 2021), whereas nanochitin has been shown to inhibit lipid digestion (Zhou et al. 2020, Zhou et al. 2021). The ability of nanofibers to reduce macronutrient digestion has been attributed to several mechanisms. First, the nanofibers may bind to key GIT components thereby altering their function, e.g., digestive enzymes, bile acids, or calcium ions. Second, the nanofibers may thicken or gel the gastrointestinal fluids, thereby retarding mixing and mass transport processes, which slows down the ability of the digestive enzymes to reach their substrates. Third, the nanofibers may form a protective coating around the macronutrients, which again decreases the ability of the digestive enzymes to access their substrates. Edible nanofibers could therefore be used as additives that retard the digestion of the macronutrients in highly processed foods, which could be useful for developing a new generation of foods that does not promote obesity and diabetes. Even so, more research is needed to establish whether these nanofibers are safe to consume and do not result in any unanticipated adverse health effects. For instance, *in-vitro* studies have shown that nanofibers can reduce the bioaccessibility of oil-soluble vitamins and nutraceuticals in model foods (Zhou et al. 2020, Zhou et al. 2021), which would be undesirable from a nutritional perspective. This effect has mainly been attributed to the reduction in lipid digestion in the presence of the nanofibers, which means that some of the vitamins and nutraceuticals remain inside the non-digested oil phase. In addition, the decrease in lipid digestion leads to a reduction in the number of mixed micelles available to solubilize any vitamins and nutraceuticals that are released.

4.3.2 Bioavailability

In principle, the healthiness of foods can be enhanced by including health promoting ingredients in them, including vitamins, nutraceuticals, and minerals (Yao et al. 2015, Mishra et al. 2021). The introduction of these bioactive ingredients into food and beverage products should not impact their desirable physicochemical and sensory qualities, such as their appearance, texture, mouthfeel, or shelf life. Moreover, to properly exhibit their beneficial health effects, these ingredients should be absorbed by the body in an active form after ingestion. Many researchers have investigated the utilization of nanotechnology to incorporate health promoting substances into food matrices in a bioavailable form (McClements 2018b, Nile et al. 2020). The bioavailability (BA) of a substance is determined by three main factors:

$$BA = B^* \times A^* \times T^*$$

Here, B^* is the bioaccessibility, which is the fraction of the substance released from the food matrix and solubilized in the gastrointestinal fluids, A^* is the fraction of this substance that is absorbed by the epithelium cells, and T^* is the fraction of this substance that escapes chemical or enzymatic transformation in the gastrointestinal environment. The relative importance of B^*, A^*, and T^* depends on the molecular characteristics of the substance, as well as food matrix effects. Extensive studies have been carried out to identify the key factors impacting the bioavailability of different kinds of bioactive substances. This knowledge is being used to create foods and beverages specifically designed to increase the bioavailability of the health promoting components they contain.

Recently, the bioavailability of the health promoting ingredients in foods has been enhanced using two different approaches where nanotechnology can be employed (McClements et al. 2015): (i) functional foods and (ii) excipient foods (Figure 4.7). In the functional food approach, the health promoting ingredients (typically oil-soluble vitamins or nutraceuticals) are loaded into edible nanoparticles (such as nanoemulsions,

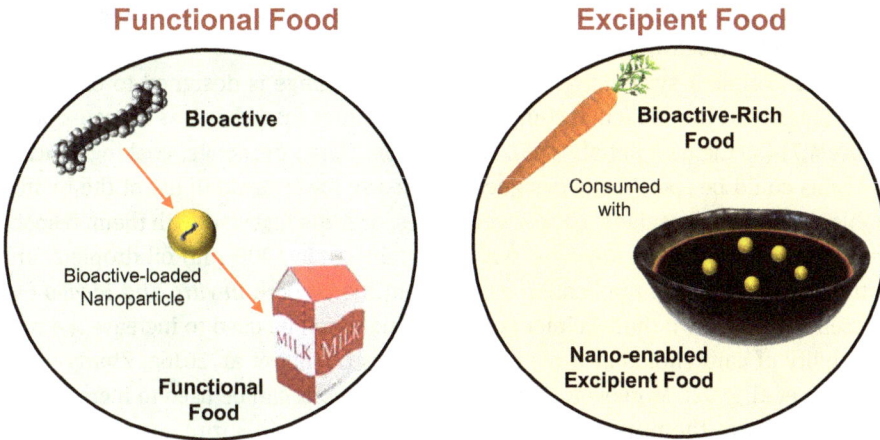

Figure 4.7: Nanotechnology can be used to increase the bioavailability of bioactive agents (like nutrients and nutraceuticals) using either functional foods or excipient foods. In functional foods, the bioactive is encapsulated within a nanoparticle that is then incorporated into a food. In excipient foods, a bioactive-rich food (like a fruit or vegetable) is consumed with a food containing nanoparticles.

microemulsions, nanoliposomes, solid lipid nanoparticles, or biopolymer nanoparticles). Encapsulation within these nanoparticles increases the water dispersibility of the oil-soluble bioactive agents, and may protect them from degradation during storage, thereby increasing the dose that is available when they are consumed. Moreover, encapsulation can increase the bioavailability of the bioactive agents in the human GIT. Lipid-based nanoparticles, like nanoemulsions or SLNs, are investigated for their potential to increase the bioavailability of oil-soluble vitamins and nutraceuticals (Yao et al. 2015, McClements 2018b). Typically, a digestible oil is used as the oil phase to formulate these delivery systems. As they pass through the stomach and small intestine, the gastric and pancreatic lipases hydrolyze these oils, which releases the bioactive agents. Moreover, the lipid digestion products, such as monoacylglycerols and free fatty acids, assemble into mixed micelles that solubilize the bioactive agents within their hydrophobic interiors. Mixed micelles are a form of natural nanoparticles formed inside the human GIT after ingestion of lipids. They consist of a mixture of micelles that are typically around a few nanometers in diameter and vesicles that are a few hundred nanometers in diameter. The hydrophobic domains in these micelles and vesicles must be large enough to accommodate the oil-soluble bioactive agents released from the lipid nanoparticles during digestion. After being solubilized within the mixed micelles, the bioactive agents are transported through the mucus layer to the lining of the small intestine where they are absorbed by the epithelium cells, packaged into chylomicrons, and then released into the lymphatic system. They then travel to the systemic circulation where they can travel around the body and be stored or utilized. Nanoparticle-based delivery systems have recently been proposed for personalized nutrition applications where several different kinds of vitamins or nutraceuticals can be incorporated into the same food or beverage system (McClements 2020b).

In the excipient system approach, a food or beverage is designed to boost the bioavailability of the bioactive components in another food that it is ingested with (Figure 4.7) (McClements et al. 2015). For example, salad dressings, cooking sauces, or creams could be specifically designed to increase the bioavailability of the hydrophobic carotenoids in salads, cooked vegetables, or fruits ingested with them, respectively. Oil-in-water nanoemulsions, which contain small (<200 nm) oil droplets, are particularly suitable for application as excipient foods. Both *in-vitro* and *in-vivo* experiments have shown that excipient nanoemulsions can be used to increase the bioavailability of carotenoids in fruits and vegetables (Zhang et al. 2016a, Zhang et al. 2019, Luo et al. 2022). Moreover, excipient nanoemulsions can be used to increase the bioavailability of the nutraceuticals or pharmaceuticals in supplements or drugs administered with them. For instance, excipient nanoemulsions have been shown to increase the bioaccessibility of carotenoids (β-carotene or lutein) in commercial carotenoid supplements consumed with them (Salvia-Trujillo and McClements 2016, Dai et al. 2021). Typically, the composition and structure of excipient nanoemulsions are designed to create an environment within the gastrointestinal tract that increases the bioavailability of the bioactive substances in foods or supplements. They may do this by increasing the solubilization of the bioactive substances within the gastrointestinal fluids (e.g., by containing digestible lipids that form mixed micelles after hydrolysis), by protecting the bioactive substances from chemical degradation (e.g., by containing antioxidants that inhibit oxidative degradation), or by increasing the fraction of the bioactive substance that is absorbed (e.g., by containing substances that increase the permeability of the epithelium cells or block efflux inhibitors) (McClements et al. 2015, Aboalnaja et al. 2016). Studies have shown that the bioaccessibility of hydrophobic nutraceuticals depends on the type and amount of oil used in the excipient nanoemulsions, as well as the oil droplet size and emulsifier type (Aboalnaja et al. 2016). For example, the bioaccessibility typically increases as the droplet size decreases because lipid digestion occurs more rapidly and completely, thereby leading to more mixed micelles being formed that can solubilize the bioactive substances released from the foods co-ingested with them (Li et al. 2017). A schematic representation of the ability of excipient nanoemulsions to increase the bioaccessibility of hydrophobic carotenoids in salad vegetables is shown in Figure 4.8. This diagram summarizes the data reported in a human feeding study that showed nanoenabled salad dressings increased the concentration of carotenoids in the bloodstream of the participants after eating salads (Luo et al. 2022).

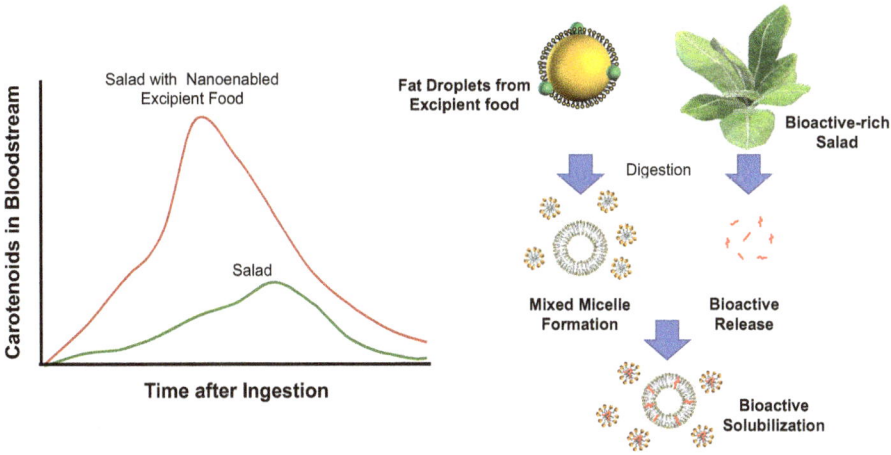

Figure 4.8: Excipient foods can be used to increase the bioavailability of bioactive agents in foods (like carotenoids in salads). For instance, nano-fat droplets rapidly digest in the gastrointestinal tract and form mixed micelles that can solubilize carotenoids released from salads, thereby increasing their bioavailability.

4.3.3 Controlled or targeted delivery of bioactives

Nanotechnology is also useful for applications where bioactive agents may need to be protected in one region of the gastrointestinal tract and then released in another region (McClements 2017). Some bioactive agents may be degraded or deactivated within the upper gastrointestinal tract (such as the mouth, stomach, or small intestine) before they reach the region where they are absorbed or can exhibit their beneficial effects (McClements 2018a). As an example, probiotics are living microorganisms that are claimed to have beneficial health effects if they can reach the colon intact and then colonize it (Sanchez et al. 2017, Kerry et al. 2018). However, many probiotics are highly susceptible to deactivation within the human stomach because of the highly acidic conditions there, as well as the presence of gastric digestive enzymes (Jiang et al. 2022). Moreover, probiotics may also be deactivated by bile salts and pancreatic digestive enzymes in the small intestine. This problem can be overcome by using nanostructured colloidal delivery systems, such as biopolymer microgels (Yao et al. 2020). For instance, studies have shown that probiotics can be encapsulated inside alginate microgels that also contain an antacid and nanoemulsion droplets (Zhang et al. 2021). The network of cross-linked alginate molecules and small oil droplets inside the microgels physically restricts the ability of digestive enzymes and bile salts to enter them, thereby protecting the probiotics. In addition, the presence of the antiacids helps to maintain neutral pH conditions inside the microgels, even when they are exposed to highly acidic gastric conditions, because some of the antacid dissolves and releases hydroxyl ions when the acids enter the microgels, thereby neutralizing them. Encapsulation of probiotics in nanostructured microgels has been shown to improve the viability and efficacy of various kinds of probiotics (Xie et al. 2021).

It may also be desirable to deliver enzymes to certain regions of the human gastrointestinal tract. For instance, people who suffer from lactose intolerance may benefit by consuming foods containing lactase, which is an enzyme that breaks down lactose so that it can be absorbed. People who suffer from pancreatitis cannot properly digest the macronutrients they consume (such as lipids) and may therefore suffer from malnutrition. Consequently, they may benefit from consuming foods containing lipase, a digestive enzyme that can hydrolyze lipids in the small intestine. However, enzymes are typically denatured and deactivated when they are exposed to the strongly acidic conditions and proteases within the stomach (McClements 2018a). This problem can also be overcome by encapsulating the enzymes within biopolymer microgels that contain antacids. Like probiotics, this kind of microgel can protect the encapsulated enzymes within the stomach and then release them within the small intestine, where they can exhibit their beneficial effects. These kinds of nanostructured microgels have been used to encapsulate, protect, and release various kinds of enzyme, including pancreatic lipase (Zhang et al. 2016b) and lactase (Zhang et al. 2016c, Zhang et al. 2017). They have also been explored for their potential for the oral delivery of insulin since they can protect this polypeptide from harsh gastric conditions (Sun et al. 2018).

4.4 Improving food safety

Commercial food products may be contaminated with spoilage or pathogenic microorganisms during their production, transport, storage, and utilization (Matthews et al. 2017). The proliferation of these microbes can raise the risks of foodborne illness and death, as well as reduce food quality and increase food waste. Microbial contamination and growth are usually controlled by adopting good manufacturing and handling practices of foods. These often involve subjecting foods to processing operations that deactivate the microbes (such as heating, dehydration, or radiation), including additives that inhibit microbial growth (such as acids and antimicrobials), and storing foods under conditions that retard microbial growth (such as freezing, chilling, or low relative humidities). Many of the preservatives currently used to control microbial growth are not label friendly or have safety concerns, e.g., benzoic acid, sulfur dioxide, and nitrates. Moreover, many of these chemical preservatives are losing their effectiveness because the microbes they are designed to treat are evolving antimicrobial resistance. For these reasons, the food industry is trying to identify natural (preferably plant-based) alternatives, which are also effective and environmentally friendly. Nanotechnology has played a key role in the creation of several kinds of antimicrobial agents intended for food applications.

Several kinds of inorganic nanoparticles exhibit antimicrobial activity, including those containing gold, silver, copper, or titanium, and can therefore be used as preservatives (Wang et al. 2017). These nanoparticles can interact with microbial cell membranes, leading to the formation of small holes that cause some of the vital substances within the microbial cells to leak out. These kinds of nanoparticles may also disrupt the functionality of the biochemical pathways inside the microbial cells by generating reactive oxygen species (ROS) that interact with key components in these pathways, such as nucleic acids, proteins, or phospholipids. Even though several kinds of inorganic nanoparticles have been shown to exhibit antimicrobial activity, they are unsuitable for direct application as preservatives in foods because they are not seen as label friendly and there are some concerns about their safety and environmental impact. Instead, they tend to be used as preservatives in packaging materials or to treat processing equipment.

Several researchers are examining the potential of nanotechnology for creating plant-based antimicrobials that can be used to inhibit the growth of pathogens and spoilage microorganisms in foods. One of the most widely studied approaches is to utilize oil-in-water nanoemulsions that contain antimicrobial essential oils isolated from plants, like cinnamon, clove, garlic, lemon, lemon grass, orange, peppermint, rosemary, tea tree, and thyme oils (Rao et al. 2019). These essential oils are naturally secreted by plants as a protection against pests that might damage them, such as microorganisms, insects, and herbivores. Essential oils contain numerous kinds of small organic molecules that exhibit antimicrobial activity, which presumably developed through evolution to tackle the wide range of pests in their environments. Studies have shown that nanoemulsions containing essential oils can inactivate a broad range of microbes by disrupting their cell membranes and critical intracellular pathways. Food-grade antimicrobial nanoemulsions have been shown to decrease bacterial contamination on produce like alfalfa seeds, mung beans, and radish seeds (Landry et al. 2015). These plant-based antimicrobial nanoemulsions may be used in the food industry in the future as a replacement for synthetic preservatives. However, it will be important that they do not adversely affect the flavor profile of foods and beverages (many of them have strong aromas), as well as maintaining their antimicrobial efficacy when introduced into complex food matrices and throughout storage.

Nanotechnology has also been utilized to produce antimicrobial treatments for foods that are based on converting water into a mist containing tiny water droplets (Huang et al. 2019). This mist is produced using an electrospray/ionization approach that generates nanosized water droplets that contain reactive oxygen species inside. These ROS have strong antimicrobial activity because they can disrupt microbial cell walls and interfere with critical biochemical pathways inside the microbes. This antimicrobial mist has been used to deactivate spoilage and pathogenic microorganisms on fruits and vegetables. This is achieved by having the fresh produce move along a conveyer belt through the mist. The antimicrobial activity of the water droplets in the mist can be enhanced by including other kinds of antimicrobial substances in the water, like hydrogen peroxide, lysozyme, nisin, and/or citric acid (Vaze et al. 2018, Huang et al. 2019, Vaze et al. 2019, Huang et al. 2021, Cohen et al. 2022).

Antimicrobial nanoparticles can be applied to food products in several ways: they can be used as additives that are incorporated into foods or beverages; they can be sprayed onto the surfaces of foods; they can be included in solutions used to wash foods; they can be included in solutions into which foods are dipped; they can be applied to the surfaces of food processing equipment; they can be included in food packaging materials. It should be noted that some of the antimicrobial nanoparticles used to clean foods, decontaminate processing equipment, or embedded within packaging materials may get into the final food products through diffusion, and thus be consumed. There are some concerns about intentionally or unintentionally including antimicrobial nanoparticles in food products. For instance, ingested antimicrobial nanoparticles may travel through the upper gastrointestinal tract and alter the microbial community residing in the colon, which could negatively influence human health and wellbeing by altering the gut microbiome (Lamas et al. 2020). For this reason, it is critical to assess the safety and environmental impacts of any newly developed antimicrobial nanoparticles before their widespread application in foods. This is true for both inorganic and organic nanoparticles.

4.5 Examples of nanostructured foods

In this section, several examples where an understanding of the nanostructure of foods is important are provided to illustrate the importance of applying nanotechnology principles within the food industry.

4.5.1 Soft drinks and other beverages

Many kinds of soft drinks and other beverages are oil-in-water nanoemulsions consisting of small oil droplets (100–500 nm) dispersed within an aqueous medium (Piorkowski and McClements 2014). The oil droplets are coated by a layer of emulsifier molecules, which facilitates their formation during homogenization and improves their stability during storage. A variety of emulsifiers can be used for this purpose, including small-molecule surfactants (like Tweens or saponins) or amphiphilic biopolymers (like gum arabic or modified starch). The oil droplets may serve different purposes. They may be carriers of oil-soluble flavors, colors, nutraceuticals, or nutrients (such as omega-3 fatty acids or vitamins), or they may be designed to provide a cloudy appearance. In the latter case, the oil droplets should have dimensions that are similar to the wavelength of light, so they scatter light strongly and make the beverage appear turbid (Figure 2.4). The oil droplets must be designed to be resistant to gravitational separation throughout the shelf life of the product (McClements 2015b). This is usually achieved by ensuring they are sufficiently small ($d < 300$ nm) and by reducing the density contrast between the oil and water phases (Section 4.2.4.1). The size of the droplets is controlled by manipulating the type and amount of emulsifier used, as well as the type and operating conditions of the homogenizer used. The density contrast between the oil and water phases is typically reduced by adding weighting agents, which are dense hydrophobic substances, to the oil phase prior to homogenization. It is also important to inhibit aggregation of the oil droplets within the beverages, as this would lead to adverse changes in the appearance and shelf life of the product. This can be achieved by ensuring that the emulsifier used can generate strong repulsive forces (usually steric and/or electrostatic) between the droplets (Section 4.2.4.2). Moreover, for flavor oils, which are relatively polar and therefore water soluble, it is important to include a ripening inhibitor to prevent the growth of the droplets due to Ostwald ripening (Section 4.2.4.3). The appearance of the final product depends on the size and concentration of oil droplets present. For beverages that should appear clear, it is important to use oil droplets with diameters below about 50 nm, but for beverages that should appear cloudy, it is better to use oil droplets with diameters between about 200 and 300 nm (Section 4.2.1).

4.5.2 Milk and milk analogs

The milk extracted from cows is a natural colloidal dispersion containing small milk fat globules that are a few micrometers big and casein micelles that are a few hundred nanometers big (Jukkola and Rojas 2017). Casein micelles are therefore a natural kind of edible nanoparticle that has been consumed for millennia. Moreover, the milk fat globules in raw milk consist of a lipid-rich core surrounded by a complex nanostructured membrane consisting of three layers of phospholipids with proteins and other substances embedded within it (Lopez et al. 2015, Verma et al. 2019). When raw milk is homogenized, the milk fat globules are broken down into fat droplets with diameters of a few hundred nanometers, which can be considered to be another kind of nanoparticle (Lopez et al. 2015). The small size of the fat droplets in homogenized milk is important because it reduces their tendency to move upwards due to gravitational forces, thereby extending the shelf life (Obeid et al. 2019).

Casein micelles are nanoparticles that typically have diameters that range between about 50 and 500 nm, which are assembled from four main kinds of protein (α_{S1}, α_{S2}, β, and k casein) and calcium phosphate nanoclusters, which are held together by a combination of salt bridges and hydrophobic attractive forces (Lucey and Horne 2018). They have a tendency to aggregate with their neighbors when the pH is adjusted close to their isoelectric point (pI ≈ 4.6) due to a decrease in the electrostatic repulsive forces acting between them (Obeid et al. 2019). They may also aggregate with each other when the polar glycopeptide portion of the kappa-casein molecules (which are normally located at the exterior of the casein micelles) is hydrolyzed by the enzyme rennet as part of the cheese manufacturing process. In this case, the removal of the hydrophilic glycopeptide decreases the steric repulsive forces acting between different casein micelles. Cow's milk is a highly versatile colloidal material that can be converted into a variety of other products through various processes, including creams, ice cream, whipped cream, butter, yogurt, and cheese. The milky appearance and creamy mouthfeel of milk and many other dairy products is a result of the ability of the nanoparticles they contain to scatter light and modify fluid flow profiles. Knowledge of the origin of these effects is important when designing plant-based analogs of real milk.

There has been growing interest in the development of plant-based alternatives to cow's milk over the past few years due to consumer concerns about the impact of the modern food supply on animal welfare, human health, and the environment (Sethi et al. 2016, McClements et al. 2019, McClements 2020a). These products are often designed to simulate the desirable appearance, texture, flavor, and nutritional profile of cow's milk, which requires a good understanding of the relationship between the composition, structure, and physicochemical properties of these systems. Nanotechnology is playing an important role in the development of the next generation of high-quality milk analogs. In general, plant-based milks can be produced using two different approaches:

- *Tissue-disruption approach*: In this top-down approach, plant tissue, such as soybeans, almonds, oats, or coconut flesh, is typically broken down into smaller fragments using mechanical and/or enzymatic approaches. For instance, the plant tissue may be ground in a mill or blended using a high shear device. The degree of tissue disruption is important to creating a final product with desirable quality attributes. Typically, the colloidal particles in the final product should have dimensions below some critical value to prevent them from feeling grainy on the tongue, to give a desirable creamy appearance, and to inhibit gravitational separation. Ideally, the tissue should be broken down until the particulate matter remaining has dimensions of a few hundred nanometers or less.
- *Homogenization approach*: In this bottom-up approach, a plant-based oil, emulsifier, and water are blended to form an oil-in-water nanoemulsion containing fat droplets in the range of a few hundred nanometers. Again, the size of these fat droplets must be controlled to obtain a desirable appearance, texture, mouthfeel, and shelf life in the final product. Moreover, the emulsifier used to form the coatings around the fat droplets must be carefully selected to ensure there is a strong repulsion between them under the conditions of use, otherwise they will tend to aggregate with each other, which may promote phase separation. This is often achieved by using protein-based emulsifiers and ensuring that the pH is well above the isoelectric point of the proteins so that they have a high negative charge. As a result, there is a strong electrostatic repulsion between

the protein-coated fat droplets, which inhibits their aggregation. Nevertheless, droplet aggregation can occur in plant-based milks when the pH becomes close to the isoelectric point of the proteins, which can occur when they are added to hot acidic coffee. The design of high-quality plant-based milks therefore depends on controlling the dimensions of the fat droplets, as well as the properties of the nanoscale interfacial layer around them.

4.5.3 Chocolate

The consistent production of high-quality chocolate depends on understanding and controlling the properties of the nanocrystalline fat that contributes to the desirable appearance, texture, and mouthfeel (Delbaere et al. 2016). The fat is organized into a network of nanocrystals assembled into complex hierarchical structures that contribute to the solid properties of chocolate, such as the characteristic "snap" of good quality products (Figure 4.9). Overall, milk chocolate is comprised of powdered milk, cocoa, and sugar dispersed within this fat crystal network. Heating causes the fat crystals to melt and disintegrate, thereby causing the chocolate to become more fluid-like. This is an important factor contributing to the desirable mouthfeel of chocolate products. The chocolate manufacturing process must be carefully regulated to create a fat crystal network that provides the desired textural attributes and mouthfeel qualities. Specifically, the dimensions, shape, polymorphic form, and interactions of the fat crystals formed within the chocolate must be controlled, which is achieved by manipulating product composition and processing conditions, such as holding temperatures, cooling rates, and shearing speeds. Researchers have studied the organization of the different structural elements in chocolate from the nanoscale to macroscale using a variety of analytical tools, including atomic force, electron, and optical microscopies. These studies have shown that small individual fat crystals ("crystallites") are assembled into star-like clusters

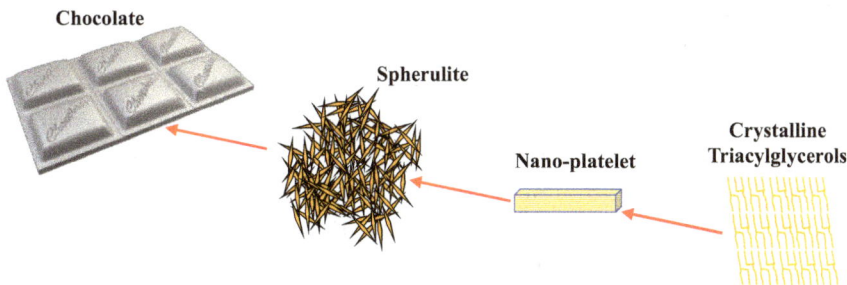

Figure 4.9: The physicochemical and sensory attributes of foods containing semi-solids fats like chocolate or butter depend on the network of nanostructured fat crystals they contain, which have a hierarchical structural organization.

("primary particles"), which are themselves assembled into larger fractal clusters ("blobs"), that aggregate with each other to provide the overall fat crystal network (Tang and Marangoni 2006b, Tang and Marangoni 2006a). Mathematical models have been developed to predict the influence of fat crystal concentration, size, and interactions on the mechanical properties of the nanostructured fat crystal networks (Marangoni and Tang 2008).

An understanding of the formation and properties of fat crystal networks is important for various other kinds of fatty foods, including butter, margarine, shortenings, and cheese. In each of these products, the structural organization of the fat crystal network impacts their texture and other quality attributes and therefore needs to be carefully controlled.

4.5.4 Encapsulation and delivery systems

One of the most widely studied applications of nanotechnology in the food industry over the past decade or so has been for the development of delivery systems to encapsulate, protect, and release bioactive components (McClements 2020b). Several of these systems were discussed earlier for the encapsulation, protection, and delivery of nutrients, nutraceuticals, probiotics, and enzymes in the gastrointestinal tract. However, the application of nanoparticles has been much broader than that. In this section, we provide more details about some of the most important uses of nanoenabled delivery systems in the food industry. A broad range of different kinds of nanoparticles have been utilized for this purpose, including microemulsions, nanoemulsions, nanoliposomes, solid lipid nanoparticles, and biopolymer nanoparticles (Chapter 2).

– *Flavors and colors*: Non-polar flavors and colors that have a low solubility in water, such as flavor oils, curcumin, and carotenoids, are often incorporated into foods and beverages in the form of nanosized lipid droplets that enhance their water dispersibility (Momin et al. 2013, de Boer et al. 2019). This kind of nanoparticle has a hydrophobic interior that can contain the non-polar flavors or colors, as well as a hydrophilic exterior that allows them to be dispersed in water. These nanoparticles may also be designed to increase the stability of the flavors and colors during storage and processing, e.g., by including preservatives such as antioxidants within them. Commonly, these non-polar substances are encapsulated within oil-in-water nanoemulsions consisting of small oil droplets dispersed in water. Some flavor oils have a finite solubility in water (such as lemon, orange, or lime oils), which means that the nanoemulsions are unstable to Ostwald ripening. These nanoemulsions must therefore be carefully formulated to inhibit droplet growth due to Ostwald ripening, which can be achieved by adding a ripening inhibitor to the oil phase (Section 4.2.3.3). A ripening inhibitor is a strongly hydrophobic substance with very low water solubility, such as edible oils containing long-chain triglycerides, like corn, sunflower, or soybean oil (Chang et al. 2012, Ryu et al. 2018). In some cases, flavors and colors need to be incorporated into clear beverages or foods without reducing their clarity. In this case, very small nanoparticles (<50 nm) that do not scatter light strongly should be used, such as those found in micelles, microemulsions, or some nanoemulsions. In other cases, the flavors and colors may need to be incorporated into products that are turbid or cloudy. In this case, larger nanoparticles (100–500 nm) that scatter light more strongly can also be used.

- *Preservatives*: Non-polar preservatives, such as antimicrobials or antioxidants, can also be encapsulated within nanosized particles to improve their water dispersibility, stability, and efficacy (Pisoschi et al. 2018). The most common example of this application is the utilization of nanoemulsions to encapsulate antimicrobial and antioxidant essential oils, such as cinnamon, rosemary, thyme, clove, lemongrass, or orange oils (Donsi and Ferrari 2016, Prakash et al. 2018, Barradas and Silva 2021). Like flavor oils, many essential oils have a finite solubility in water, which makes nanoemulsions prepared from them highly unstable to droplet growth due to Ostwald ripening. Again, the stability of these nanoemulsions can be improved by incorporating a suitable ripening inhibitor in the oil phase. However, the amount used should not be too high, otherwise the preservatives may lose their efficacy because their tendency to associate with the microbial surfaces is reduced (Ryu et al. 2018). The interfacial properties of the oil droplets in the nanoemulsions can also be designed to increase their effectiveness as preservatives (Ziani et al. 2011, Wu et al. 2019). For instance, positively charged droplets tend to be attracted to negatively charged microbial cells, thereby bringing the antimicrobial essential oils into closer contact with the microorganisms, which may increase their efficacy. Certain kinds of phytochemicals with antimicrobial or antioxidant properties can also be solubilized within the oil phase of nanoemulsions to increase their preservative properties, such as curcumin, quercetin, or resveratrol.
- *Vitamins and nutraceuticals*: Edible nanoparticles, especially nanoemulsions, have also been widely investigated for their ability to encapsulate non-polar vitamins and nutraceuticals like vitamins A, D, and E, carotenoids, curcumin, and polyphenols (Salvia-Trujillo et al. 2017b, Choi and McClements 2020). These systems are typically designed to increase the water dispersibility, chemical stability, and bioavailability of the encapsulated substances. The nanoparticle size, composition, and interfacial properties must be carefully controlled to achieve these goals. For instance, studies have shown that the bioaccessibility and/or bioavailability of strongly hydrophobic bioactives, like oil-soluble vitamins and carotenoids, increases as the size of the oil droplets in nanoemulsions decreases (Parthasarathi et al. 2016, Salvia-Trujillo et al. 2017a). This effect has been attributed to the increase in droplet surface area, which leads to more rapid

and complete lipid digestion in the gastrointestinal tract. As a result, more of the bioactives are released from the oil droplets and solubilized in the mixed micelles in the small intestine. Studies have also shown that the nature of the oil phase used in the nanoemulsions influences the bioaccessibility and/or bioavailability of any encapsulated hydrophobic bioactives (Salvia-Trujillo et al. 2015, Salvia-Trujillo and McClements 2016, Yang et al. 2017). Typically, an oil phase should be used that when digested in the small intestine it produces mixed micelles containing hydrophobic domains large enough to accommodate the bioactives inside them. The size of these hydrophobic domains depends on the chain length and degree of unsaturation of the fatty acid chains in the oil phase (McClements 2018b). For example, many studies have shown that carotenoids have a much higher bioaccessibility when they are dissolved in an oil phase containing long-chain fatty acids (such as corn oil) than when they are dissolved in one containing medium- or short-chain fatty acids (such as MCT or coconut oil) (Ozturk et al. 2015, Zhang et al. 2015). Nanoemulsions may also be designed to inhibit the chemical degradation of vitamins and nutraceuticals during storage, e.g., by incorporating antioxidants or chelating agents, or by modifying the interfacial properties (Mao et al. 2010, Qian et al. 2012a, Qian et al. 2012b).

– *Probiotics and enzymes*: As mentioned earlier, probiotics and enzymes can be encapsulated within biopolymer microgels to improve their resistance to degradation during storage and passage through the gastrointestinal tract. In the case of probiotics, nanoparticles cannot be directly used for this purpose because the microbes themselves are typically a few micrometers big. Even so, the nanostructure of the microgel interior can be manipulated to improve the stability and release characteristics of encapsulated probiotics and enzymes. For instance, the size of the pores within the microgels can be altered by changing the concentrations of the biopolymers or crosslinking agents used, or by adding solid or liquid nanoparticles to fill in the pores (Yao et al. 2018, Yao et al. 2020). Reducing the pore size inhibits the ability of digestive enzymes and bile salts to enter the microgels and deactivate the probiotics.

4.6 Conclusions

Nanotechnology has already had a major impact on improving the quality, nutrition, shelf life, and safety of food and beverage products. The most common application has been the utilization of nanoparticles to encapsulate, protect, and deliver bioactive agents, including macronutrients, vitamins, nutraceuticals, colors, flavors, and preservatives. Some of these nanoenabled systems are now finding utilization in commercial supplements and food products. For instance, they are being used to increase the bioavailability of oil-soluble vitamins and nutraceuticals in functional foods and beverages. Nanotechnology can also be used to create "special effects" in foods, such as controlled or triggered release profiles, which may be desirable for certain applications. It is likely that nanotechnology will find increasing application in the food industry. However, it will be important to ensure that any nanomaterials used in foods are safe to consume. Moreover, it will be important to inform consumers about the potential risks and benefits of consuming nano-foods so they can make informed choices about what they eat.

References

Aboalnaja, K. O., S. Yaghmoor, T. A. Kumosani and D. J. McClements (2016). "Utilization of nanoemulsions to enhance bioactivity of pharmaceuticals, supplements, and nutraceuticals: Nanoemulsion delivery systems and nanoemulsion excipient systems". Expert Opinion on Drug Delivery **13**(9): 1327–1336.

Akhavan, S., E. Assadpour, I. Katouzian and S. M. Jafari (2018). "Lipid nano scale cargos for the protection and delivery of food bioactive ingredients and nutraceuticals". Trends in Food Science & Technology **74**: 132–146.

Barradas, T. N. and K. Silva (2021). "Nanoemulsions of essential oils to improve solubility, stability and permeability: a review". Environmental Chemistry Letters **19**(2): 1153–1171.

Chang, Y. H., L. McLandsborough and D. J. McClements (2012). "Physical properties and antimicrobial efficacy of thyme oil nanoemulsions: Influence of ripening inhibitors". Journal of Agricultural and Food Chemistry **60**(48): 12056–12063.

Choi, S. J. and D. J. McClements (2020). "Nanoemulsions as delivery systems for lipophilic nutraceuticals: strategies for improving their formulation, stability, functionality and bioavailability". Food Science and Biotechnology **29**(2): 149–168.

Cohen, Y., E. Mwangi, N. Tish, J. Xu, N. D. Vaze, T. Klingbell, E. Fallik, Y. G. Luo, P. Demokritou, V. Rodov and E. Poverenov (2022). Quaternized chitosan as a biopolymer sanitizer for leafy vegetables: synthesis, characteristics, and traditional vs. dry nano-aerosol applications. Food Chemistry **378**: 132056, 1–10. doi.org/10.1016/j.foodchem.2022.132056

Dai, L., L. Y. Zhou, H. L. Zhou, B. J. Zheng, N. Ji, X. F. Xu, X. Y. He, L. Xiong, D. J. McClements and Q. J. Sun (2021). "Comparison of lutein bioaccessibility from dietary supplement-excipient nanoemulsions and nanoemulsion-based delivery systems". Journal of Agricultural and Food Chemistry **69**(46): 13925–13932.

de Boer, F. Y., A. Imhof and K. P. Velikov (2019). "Encapsulation of colorants by natural polymers for food applications". Coloration Technology **135**(3): 183–194.

Delbaere, C., D. Van de Walle, F. Depypere, X. Gellynck and K. Dewettinck (2016). "Relationship between chocolate microstructure, oil migration, and fat bloom in filled chocolates". European Journal of Lipid Science and Technology **118**(12): 1800–1826.

DeLoid, G. M., I. S. Sohal, L. R. Lorente, R. M. Molina, G. Pyrgiotakis, A. Stevanovic, R. Zhang, D. J. McClements, N. K. Geitner, D. W. Bousfield, K. W. Ng, S. C. J. Loo, D. C. Bell, J. Brain and P. Demokritou (2018a). "Reducing intestinal digestion and absorption of fat using a nature-derived biopolymer: interference of triglyceride hydrolysis by nanocellulose". ACS nano **12**(7): 6469–6479.

DeLoid, G. M., I. S. Sohal, L. R. Lorente, R. M. Molina, G. Pyrgiotakis, A. Stevanovic, R. J. Zhang, D. J. McClements, N. K. Geitner, D. W. Bousfield, K. W. Ng, S. C. J. Loo, D. C. Bell, J. Brain and P. Demokritou (2018b). "Reducing intestinal digestion and absorption of fat using a

nature-derived biopolymer: interference of triglyceride hydrolysis by nanocellulose".
Acs Nano **12**(7): 6469–6479.

Donsi, F. and G. Ferrari (2016). "Essential oil nanoemulsions as antimicrobial agents in food".
Journal of Biotechnology **233**: 106–120.

Fredrick, E., P. Walstra and K. Dewettinck (2010). "Factors governing partial coalescence in oil-in-water emulsions". Advances in Colloid and Interface Science **153**(1–2): 30–42.

Guo, Z. Y., G. M. DeLoid, X. Q. Cao, D. Bitounis, K. Sampathkumar, K. W. Ng, S. C. J. Loo and
P. Demokritou (2021). "Effects of ingested nanocellulose and nanochitosan materials on carbohydrate digestion and absorption in an in vitro small intestinal epithelium model".
Environmental Science-Nano **8**(9): 2554–2568.

Huang, R. Z., N. Vaze, A. Soorneedi, M. D. Moore, Y. G. Luo, E. Poverenov, V. Rodov and
P. Demokritou (2021). "A novel antimicrobial technology to enhance food safety and quality of leafy vegetables using engineered water nanostructures". Environmental Science-Nano **8**(2):
514–526.

Huang, R. Z., N. Vaze, A. Soorneedi, M. D. Moore, Y. L. Xue, D. Bello and P. Demokritou (2019).
"Inactivation of hand hygiene-related pathogens using engineered water nanostructures". Acs Sustainable Chemistry & Engineering **7**(24): 19761–19769.

Hutchings, J. B. (1999). Food Color and Appearance. Gaithersburg, MD, Aspen Publishers.

Ikeda, S. and Q. X. Zhong (2012). "Polymer and colloidal models describing structure-function relationships". Annual Review of Food Science and Technology Vol 3. M. P. Doyle and
T. R. Klaenhammer **3**: 405–424.

Israelachvili, J. (2011). Intermolecular and Surface Forces. Third Edition. London, UK, Academic Press.

Jiang, Z. W., M. T. Li, D. J. McClements, X. B. Liu and F. G. Liu (2022). Recent advances in the design and fabrication of probiotic delivery systems to target intestinal inflammation. Food Hydrocolloids **125**: 107438, 1–9. doi.org/10.1016/j.foodhyd.2021.107438

Jukkola, A. and O. J. Rojas (2017). "Milk fat globules and associated membranes: Colloidal properties and processing effects". Advances in Colloid and Interface Science **245**: 92–101.

Kabalnov, A. (2001). "Ostwald ripening and related phenomena". Journal of Dispersion Science and Technology **22**(1): 1–12.

Kabalnov, A. S. and E. D. Shchukin (1992). "Ostwald ripening theory – applications to fluorocarbon emulsion stability". Advances in Colloid and Interface Science **38**: 69–97.

Kerry, R. G., J. K. Patra, S. Gouda, Y. Park, H. S. Shin and G. Das (2018). "Benefaction of probiotics for human health: A review". Journal of Food and Drug Analysis **26**(3): 927–939.

Koroleva, M. Y. and E. V. Yurtov (2021). "Ostwald ripening in macro- and nanoemulsions". Russian Chemical Reviews **90**(3): 293–323.

Lamas, B., N. M. Breyner and E. Houdeau (2020). "Impacts of foodborne inorganic nanoparticles on the gut microbiota-immune axis: potential consequences for host health". Particle and Fibre Toxicology **17**(1): 19, 1–12. doi.org/10.1186/s12989-020-00349-z.

Landry, K. S., S. Micheli, D. J. McClements and L. McLandsborough (2015). "Effectiveness of a spontaneous carvacrol nanoemulsion against Salmonella enterica Enteritidis and Escherichia coli O157: H7on contaminated broccoli and radish seeds". Food Microbiology **51**: 10–17.

Lane, M. M., J. A. Davis, S. Beattie, C. Gomez-Donoso, A. Loughman, A. O'Neil, F. Jacka, M. Berk,
R. Page, W. Marx and T. Rocks (2021). "Ultraprocessed food and chronic noncommunicable diseases: A systematic review and meta-analysis of 43 observational studies". Obesity Reviews **22**(3): e13146, 1–15. doi.org/10.1111/obr.13146.

Larson, R. G. (1999). The Structure and Rheology of Complex Fluids. Oxford, UK, Oxford, Oxford University Press.

Li, Q., T. Li, C. M. Liu, T. T. Dai, R. J. Zhang, Z. P. Zhang and D. J. McClements (2017). "Enhancement of Carotenoid Bioaccessibility from Tomatoes Using Excipient Emulsions: Influence of Particle Size". Food Biophysics 12(2): 172–185.

Li, Y., S. Le Maux, H. Xiao and D. J. McClements (2009). "Emulsion-Based Delivery Systems for Tributyrin, a Potential Colon Cancer Preventative Agent". Journal of Agricultural and Food Chemistry 57(19): 9243–9249.

Liu, L. L. and F. B. Kong (2019). "In vitro investigation of the influence of nano-cellulose on starch and milk digestion and mineral adsorption". International Journal of Biological Macromolecules 137: 1278–1285.

Liu, L. L. and F. B. Kong (2021). The behavior of nanocellulose in gastrointestinal tract and its influence on food digestion. Journal of Food Engineering 292: 110346, 1–13. doi.org/10.1016/j.jfoodeng.2020.110346.

Lopez, C., C. Cauty and F. Guyomarc'h (2015). "Organization of lipids in milks, infant milk formulas and various dairy products: role of technological processes and potential impacts". Dairy Science & Technology 95(6): 863–893.

Lucey, J. A. and D. S. Horne (2018). "Perspectives on casein interactions". International Dairy Journal 85: 56–65.

Luo, H., Z. Li, C. R. Straight, Q. Wang, J. Zhou, Y. Sun, C.-Y. Lo, L. Yi, Y. Wu, J. Huang, W. Wolfe, D. Z. Sutherland, M. S. Miller, D. J. McClements, E. A. Decker and H. Xiao (2022). "Black pepper and vegetable oil-based emulsion synergistically enhance carotenoid bioavailability of raw vegetables in humans". Food Chemistry 373: 131277, 1–9. doi.org/https://doi.org/10.1016/j.foodchem.2021.131277.

Mao, L. K., J. Yang, D. X. Xu, F. Yuan and Y. X. Gao (2010). "Effects of Homogenization Models and Emulsifiers on the Physicochemical Properties of -Carotene Nanoemulsions". Journal of Dispersion Science and Technology 31(7): 986–993.

Mao, Y. Y. and D. J. McClements (2012). "Modulation of emulsion rheology through electrostatic heteroaggregation of oppositely charged lipid droplets: Influence of particle size and emulsifier content". Journal of Colloid and Interface Science 380: 60–66.

Marangoni, A. G. and D. M. Tang (2008). "Modeling the rheological properties of fats: A perspective and recent advances". Food Biophysics 3(2): 113–119.

Matthews, K. R., K. E. Kniel and T. J. Montville (2017). Food Microbiology: An Introduction. Washington, D.C., ASM Press.

McClements, D. J. (2002a). "Colloidal basis of emulsion color". Current Opinion in Colloid & Interface Science 7(5–6): 451–455.

McClements, D. J. (2002b). "Theoretical prediction of emulsion color". Advances in Colloid and Interface Science 97(1–3): 63–89.

McClements, D. J. (2015a). Food Emulsions: Principles, Practice, and Techniques. Boca Raton, CRC Press.

McClements, D. J. (2015b). Food Emulsions: Principles, Practice, and Techniques. Boca Raton, Florida, US, CRC Press.

McClements, D. J. (2015c). Nanoparticle- and Microparticle-based Delivery Systems. Boca Raton, FL, CRC Press.

McClements, D. J. (2017). "Designing biopolymer microgels to encapsulate, protect and deliver bioactive components: Physicochemical aspects". Advances in Colloid and Interface Science 240: 31–59.

McClements, D. J. (2018a). "Encapsulation, protection, and delivery of bioactive proteins and peptides using nanoparticle and microparticle systems: A review". Advances in Colloid and Interface Science 253: 1–22.

McClements, D. J. (2018b). "Enhanced delivery of lipophilic bioactives using emulsions: a review of major factors affecting vitamin, nutraceutical, and lipid bioaccessibility". Food & Function **9** (1): 22–41.

McClements, D. J. (2019). Future Foods: How Modern Science is Transforming the Way We Eat. Cham, Switzerland, Springer Scientific.

McClements, D. J. (2020a). "Development of Next-Generation Nutritionally Fortified Plant-Based Milk Substitutes: Structural Design Principles". Foods 9(4): 421, 1–19. doi.org/10.3390/foods9040421.

McClements, D. J. (2020b). "Nano-enabled personalized nutrition: Developing multicomponent-bioactive colloidal delivery systems". Advances in Colloid and Interface Science **282**: 102211, 1–15. doi.org/10.1016/j.cis.2020.102211.

McClements, D. J. (2020c). "Recent advances in the production and application of nano-enabled bioactive food ingredients". Current Opinion in Food Science **33**: 85–90.

McClements, D. J., E. Newman and I. F. McClements (2019). "Plant-based Milks: A Review of the Science Underpinning Their Design, Fabrication, and Performance". Comprehensive Reviews in Food Science and Food Safety **18**(6): 2047–2067.

McClements, D. J., H. Xiao and P. Demokritou (2017). "Physicochemical and colloidal aspects of food matrix effects on gastrointestinal fate of ingested inorganic nanoparticles". Advances in Colloid and Interface Science **246**: 165–180.

McClements, D. J., L. Q. Zou, R. J. Zhang, L. Salvia-Trujillo, T. Kumosani and H. Xiao (2015). "Enhancing Nutraceutical Performance Using Excipient Foods: Designing Food Structures and Compositions to Increase Bioavailability". Comprehensive Reviews in Food Science and Food Safety **14**(6): 824–847.

Mishra, A., D. Pradhan, P. Biswasroy, B. Kar, G. Ghosh and G. Rath (2021). Recent advances in colloidal technology for the improved bioavailability of the nutraceuticals. Journal of Drug Delivery Science and Technology **65**: 102693, 1–15. doi.org/10.1016/j.jddst.2021.102693.

Momin, J. K., C. Jayakumar and J. B. Prajapati (2013). "Potential of nanotechnology in functional foods". Emirates Journal of Food and Agriculture **25**(1): 10–19.

Nile, S. H., V. Baskar, D. Selvaraj, A. Nile, J. B. Xiao and G. Y. Kai (2020). "Nanotechnologies in Food Science: Applications, Recent Trends, and Future Perspectives". Nano-Micro Letters **12**(1): 45, 1–14. doi.org/10.1007/s40820-020-0383-9.

Obeid, S., F. Guyomarc'h, G. Francius, H. Guillemin, X. X. Wu, S. Pezennec, M. H. Famelart, C. Cauty, F. Gaucheron and C. Lopez (2019). The surface properties of milk fat globules govern their interactions with the caseins: Role of homogenization and pH probed by AFM force spectroscopy. Colloids and Surfaces B-Biointerfaces **182**: 110363, 1–8. doi.org/10.1016/j.colsurfb.2019.110363.

Ozturk, B., S. Argin, M. Ozilgen and D. J. McClements (2015). "Nanoemulsion delivery systems for oil-soluble vitamins: Influence of carrier oil type on lipid digestion and vitamin D-3 bioaccessibility". Food Chemistry **187**: 499–506.

Parthasarathi, S., S. P. Muthukumar and C. Anandharamakrishnan (2016). "The influence of droplet size on the stability, in vivo digestion, and oral bioavailability of vitamin E emulsions". Food & Function **7**(5): 2294–2302.

Petrut, R. F., S. Danthine and C. Blecker (2016). "Assessment of partial coalescence in whippable oil-in-water food emulsions". Advances in Colloid and Interface Science **229**: 25–33.

Piorkowski, D. T. and D. J. McClements (2014). "Beverage emulsions: Recent developments in formulation, production, and applications". Food Hydrocolloids **42**: 5–41.

Pisoschi, A. M., A. Pop, C. Cimpeanu, V. Turcus, G. Predoi and F. Iordache (2018). "Nanoencapsulation techniques for compounds and products with antioxidant and antimicrobial activity – A critical view". European Journal of Medicinal Chemistry **157**: 1326–1345.

Prakash, A., R. Baskaran, N. Paramasivam and V. Vadivel (2018). "Essential oil based nanoemulsions to improve the microbial quality of minimally processed fruits and vegetables: A review". Food Research International **111**: 509–523.

Qian, C., E. A. Decker, H. Xiao and D. J. McClements (2012a). "Inhibition of beta-carotene degradation in oil-in-water nanoemulsions: Influence of oil-soluble and water-soluble antioxidants". Food Chemistry **135**(3): 1036–1043.

Qian, C., E. A. Decker, H. Xiao and D. J. McClements (2012b). "Physical and chemical stability of beta-carotene-enriched nanoemulsions: Influence of pH, ionic strength, temperature, and emulsifier type". Food Chemistry **132**(3): 1221–1229.

Rao, J. J., B. C. Chen and D. J. McClements (2019). "Improving the Efficacy of Essential Oils as Antimicrobials in Foods: Mechanisms of Action". Annual Review of Food Science and Technology Vol 10. M. P. Doyle and D. J. McClements **10**: 365–387.

Ryu, V., D. J. McClements, M. G. Corradini and L. McLandsborough (2018). "Effect of ripening inhibitor type on formation, stability, and antimicrobial activity of thyme oil nanoemulsion". Food Chemistry **245**: 104–111.

Salvia-Trujillo, L., B. Fumiaki, Y. Park and D. J. McClements (2017a). "The influence of lipid droplet size on the oral bioavailability of vitamin D-2 encapsulated in emulsions: an in vitro and in vivo study". Food & Function **8**(2): 767–777.

Salvia-Trujillo, L. and D. J. McClements (2016). "Improvement of beta-Carotene Bioaccessibility from Dietary Supplements Using Excipient Nanoemulsions". Journal of Agricultural and Food Chemistry **64**(22): 4639–4647.

Salvia-Trujillo, L., R. Soliva-Fortuny, M. A. Rojas-Grau, D. J. McClements and O. Martin-Belloso (2017b). "Edible Nanoemulsions as Carriers of Active Ingredients: A Review". Annual Review of Food Science and Technology Vol 8. M. P. Doyle and T. R. Klaenhammer **8**: 439–466.

Salvia-Trujillo, L., Q. Sun, B. H. Urn, Y. Park and D. J. McClements (2015). "In vitro and in vivo study of fucoxanthin bioavailability from nanoemulsion-based delivery systems: Impact of lipid carrier type". Journal of Functional Foods **17**: 293–304.

Sanchez, B., S. Delgado, A. Blanco-Miguez, A. Lourenco, M. Gueimonde and A. Margolles (2017). "Probiotics, gut microbiota, and their influence on host health and disease". Molecular Nutrition & Food Research **61**(1): 1600240, 1–13, doi.org/10.1002/mnfr.201600240.

Sethi, S., S. K. Tyagi and R. K. Anurag (2016). "Plant-based milk alternatives an emerging segment of functional beverages: a review". Journal of Food Science and Technology-Mysore **53**(9): 3408–3423.

Sun, Q. C., Z. P. Zhang, R. J. Zhang, R. C. Gao and D. J. McClements (2018). "Development of Functional or Medical Foods for Oral Administration of Insulin for Diabetes Treatment: Gastroprotective Edible Microgels". Journal of Agricultural and Food Chemistry **66**(19): 4820–4826.

Tang, D. M. and A. G. Marangoni (2006a). "Microstructure and fractal analysis of fat crystal networks". Journal of the American Oil Chemists Society **83**(5): 377–388.

Tang, D. M. and A. G. Marangoni (2006b). "Quantitative study on the microstructure of colloidal fat crystal networks and fractal dimensions". Advances in Colloid and Interface Science **128**: 257–265.

Taylor, P. (1998). "Ostwald ripening in emulsions". Advances in Colloid and Interface Science **75**(2): 107–163.

Vaze, N., Y. Jiang, L. Mena, Y. P. Zhang, D. Bello, S. S. Leonard, A. M. Morris, M. Eleftheriadou, G. Pyrgiotakis and P. Demokritou (2018). "An integrated electrolysis – electrospray –

ionization antimicrobial platform using Engineered Water Nanostructures (EWNS) for food
safety applications". Food Control **85**: 151–160.

Vaze, N., G. Pyrgiotakis, L. Mena, R. Baumann, A. Demokritou, M. Ericsson, Y. P. Zhang,
D. Bello, M. Eleftheriadou and P. Demokritou (2019). "A nano-carrier platform for the targeted
delivery of nature-inspired antimicrobials using Engineered Water Nanostructures for food
safety applications". Food Control **96**: 365–374.

Verma, A., T. Ghosh, B. Bhushan, G. Packirisamy, N. K. Navani, P. P. Sarangi and K. Ambatipudi
(2019). "Characterization of difference in structure and function of fresh and mastitic bovine
milk fat globules". Plos One **14**(8): e0221830, 1–9, doi.org/10.1371/journal.pone.0221830

Wang, L. L., C. Hu and L. Q. Shao (2017). "The antimicrobial activity of nanoparticles: present
situation and prospects for the future". International Journal of Nanomedicine **12**: 1227–1249.

Wang, M. Q., T. Doi and D. J. McClements (2019). "Encapsulation and controlled release of
hydrophobic flavors using biopolymer-based microgel delivery systems: Sustained release of
garlic flavor during simulated cooking". Food Research International **119**: 6–14.

Wooster, T. J., M. Golding and P. Sanguansri (2008). "Impact of Oil Type on Nanoemulsion
Formation and Ostwald Ripening Stability". Langmuir **24**(22): 12758–12765.

Wu, D. H., J. Lu, S. B. Zhong, P. Schwarz, B. C. Chen and J. J. Rao (2019). "Influence of nonionic and
ionic surfactants on the antifungal and mycotoxin inhibitory efficacy of cinnamon oil
nanoemulsions". Food & Function **10**(5): 2817–2827.

Xie, J. J., M. F. Yao, Y. M. Lu, M. J. Yu, S. Y. Han, D. J. McClements, H. Xiao and L. J. Li (2021). "Impact
of encapsulating a probiotic (Pediococcus pentosaceus Li05) within gastro-responsive
microgels on Clostridium difficile infections". Food & Function **12**(7): 3180–3190.

Yang, Y., H. Xiao and D. J. McClements (2017). "Impact of Lipid Phase on the Bioavailability of
Vitamin E in Emulsion-Based Delivery Systems: Relative Importance of Bioaccessibility,
Absorption, and Transformation". Journal of Agricultural and Food Chemistry **65**(19):
3946–3955.

Yao, M. F., B. Li, H. W. Ye, W. H. Huang, Q. X. Luo, H. Xiao, D. J. McClements and L. J. Li (2018).
"Enhanced viability of probiotics (Pediococcus pentosaceus Li05) by encapsulation in
microgels doped with inorganic nanoparticles". Food Hydrocolloids **83**: 246–252.

Yao, M. F., D. J. McClements and H. Xiao (2015). "Improving oral bioavailability of nutraceuticals by
engineered nanoparticle-based delivery systems". Current Opinion in Food Science **2**: 14–19.

Yao, M. F., J. J. Xie, H. J. Du, D. J. McClements, H. Xiao and L. J. Li (2020). "Progress in
microencapsulation of probiotics: A review". Comprehensive Reviews in Food Science and
Food Safety **19**(2): 857–874.

Zhang, R. J., W. H. Wu, Z. P. Zhang, S. S. Lv, B. S. Xing and D. J. McClements (2019). "Impact of Food
Emulsions on the Bioaccessibility of Hydrophobic Pesticide Residues in Co-Ingested Natural
Products: Influence of Emulsifier and Dietary Fiber Type". Journal of Agricultural and Food
Chemistry **67**(21): 6032–6040.

Zhang, R. J., Z. P. Zhang, L. Q. Zou, H. Xiao, G. D. Zhang, E. A. Decker and D. J. McClements (2015).
"Enhancing Nutraceutical Bioavailability from Raw and Cooked Vegetables Using Excipient
Emulsions: Influence of Lipid Type on Carotenoid Bioaccessibility from Carrots". Journal of
Agricultural and Food Chemistry **63**(48): 10508–10517.

Zhang, R. J., Z. P. Zhang, L. Q. Zou, H. Xiao, G. D. Zhang, E. A. Decker and D. J. McClements (2016a).
"Impact of Lipid Content on the Ability of Excipient Emulsions to Increase Carotenoid
Bioaccessibility from Natural Sources (Raw and Cooked Carrots)". Food Biophysics **11**(1):
71–80.

Zhang, Z. P., F. Chen, R. J. Zhang, Z. Y. Deng and D. J. McClements (2016b). "Encapsulation of
Pancreatic Lipase in Hydrogel Beads with Self-Regulating Internal pH Microenvironments:

Retention of Lipase Activity after Exposure to Gastric Conditions". Journal of Agricultural and Food Chemistry **64**(51): 9616–9623.

Zhang, Z. P., M. Gu, X. M. You, D. A. Sela, H. Xiao and D. J. McClements (2021). Encapsulation of bifidobacterium in alginate microgels improves viability and targeted gut release. Food Hydrocolloids **116**: 106634, 1–11. doi.org/10.1016/j.foodhyd.2021.106634

Zhang, Z. P., R. J. Zhang, L. Chen and D. J. McClements (2016c). "Encapsulation of lactase (beta-galactosidase) into kappa-carrageenan-based hydrogel beads: Impact of environmental conditions on enzyme activity". Food Chemistry **200**: 69–75.

Zhang, Z. P., R. J. Zhang and D. J. McClements (2017). "Lactase (beta-galactosidase) encapsulation in hydrogel beads with controlled internal pH microenvironments: Impact of bead characteristics on enzyme activity". Food Hydrocolloids **67**: 85–93.

Zhou, H. L., T. T. Dai, J. N. Liu, Y. B. Tan, L. Bai, O. J. Rojas and D. J. McClements (2021). Chitin nanocrystals reduce lipid digestion and beta-carotene bioaccessibility: An in-vitro INFOGEST gastrointestinal study. Food Hydrocolloids **113**: 106494, 1–12. doi.org/10.1016/j. foodhyd.2020.106494

Zhou, H. L., Y. B. Tan, S. S. Lv, J. N. Liu, J. L. M. Mundo, L. Bai, O. J. Rojas and D. J. McClements (2020). Nanochitin-stabilized pickering emulsions: Influence of nanochitin on lipid digestibility and vitamin bioaccessibility. Food Hydrocolloids **106**: 105878, 1–12. doi.org/ 10.1016/j.foodhyd.2020.105878

Ziani, K., Y. H. Chang, L. McLandsborough and D. J. McClements (2011). "Influence of Surfactant Charge on Antimicrobial Efficacy of Surfactant-Stabilized Thyme Oil Nanoemulsions". Journal of Agricultural and Food Chemistry **59**(11): 6247–6255.

Chapter 5
Advanced nanomaterials for food and agriculture applications

5.1 Introduction

Nanotechnology has been applied in the food and agriculture areas to create a variety of advanced nanostructured materials, such as sensors, packaging materials, and filters (Otoni et al. 2021). These advanced materials usually contain nanoscale structures that bring new or improved functional attributes. The substances used to create these advanced nanomaterials are often obtained from agricultural or food waste generated by farms, manufacturing operations, retailers, or consumers (Otoni et al. 2021). The nanomaterials created are not typically intended to be consumed but are designed to provide other important functions for either food or non-food applications. In this chapter, an overview of the use of nanotechnology to create sensors, packaging materials, filters, and other useful materials is given.

5.2 Nano-enabled sensor technologies

Knowledge of the properties of foods, such as their composition or structure, is important for determining their quality, safety, healthiness, and shelf-life. This kind of information is typically obtained using dedicated analytical instruments found in a research, development, or quality assurance laboratories, such as chromatography, spectroscopy, or mass spectrometry instruments (Nielsen 2017). These instruments are often relatively expensive to purchase and require trained personnel to operate. Moreover, sample preparation and analysis can be a time-consuming and laborious process. Consequently, there is interest in the development of new sensor technologies to provide information about food properties that are inexpensive, simple to use, rapid, and reliable. Common examples of this kind of advanced sensor technology are home pregnancy and COVID-19 test kits.

Nanotechnology can play an important role in the development of new sensor technologies for application in the food and agriculture industry (Wang and Duncan 2017). These sensors could be used by farmers, food manufacturers, distributers, retailers, and consumers to provide valuable information about the composition, freshness, quality, or safety of food and beverage products rapidly and cheaply. Nano-enabled sensor could be incorporated into simple hand-held devices or integrated into food packaging materials. Alternatively, nanotechnology can be used to increase the efficacy of existing analytical instruments, as has been achieved with Raman spectroscopy where placing

https://doi.org/10.1515/9783110788457-005

samples on nanostructured materials can greatly increase the strength of the signal obtained (Petersen et al. 2021).

A variety of nanoparticle-based sensors have been developed for application within the food and agricultural industries (Bulbul et al. 2015, Krishna et al. 2018, Liu et al. 2018, Li et al. 2019b). These sensors must have at least two elements: a detector and a transducer (Li et al. 2019b). The detection element interacts with a target substance (the analyte) in the test matrix, while the transducer element generates a signal that can be measured. The advantage of using nanomaterials as detection elements in sensors is that they have a high specific surface area, which means that multiple sensing elements can be included, thereby increasing their sensitivity. Moreover, some kinds of nanoparticles have unique physicochemical properties that allow them to be used as transducer elements, such as the optical properties of gold nanoparticles or the magnetic properties of iron oxide nanoparticles. In this section, several examples of sensor technologies that utilize nanomaterials are highlighted.

5.2.1 Nanomaterials in surface-enhanced Raman spectroscopy

Raman spectroscopy is a powerful analytical tool for providing information about the chemical constituents and molecular interactions within materials (Wang et al. 2021). It is based on the ability of electromagnetic waves with certain frequencies to stimulate the vibrations of molecular groups within materials. Whereas infrared spectroscopy relies on the absorption of energy from electromagnetic waves, Raman spectroscopy is based on the exchange of energy (Ellis et al. 2012). A Raman spectrometer directs a light beam at the material being tested and measures the loss or gain in energy of the photons (change in their frequency or wavenumber) due to their inelastic interactions with the molecules in the material. Each type of molecule has a unique set of molecular vibrations that lead to a unique spectrum, which acts as a fingerprint of the material being tested. The positions of the peaks in a Raman spectrum can be used to identify the type of chemical groups it contains, which can provide valuable information about its composition. Like infrared spectroscopy, Raman spectroscopy is also capable of rapid analysis and can determine the presence of a wide range of food components without destroying the material being examined. It can also measure the properties of foods contained in plastic or glass packages, which is an advantage for some applications. Another major advantage of Raman spectroscopy is that it is more suitable for analyzing samples with higher water contents than infrared spectroscopy. Raman spectroscopy has been used to detect and quantify a broad spectrum of substances within foods, including nutrients, colors, flavors, preservatives, microorganisms, pesticides, antibiotics, and other contaminants (Huang et al. 2020, Wang et al. 2021, Yang et al. 2021). Consequently, it can be used to provide valuable information about food quality, freshness, and safety.

One of the main drawbacks of conventional Raman spectroscopy is that it does not have a high sensitivity. However, its sensitivity can be dramatically enhanced by using nanomaterials as substrates, which is known as surface-enhanced Raman spectroscopy or SERS (Jiang et al. 2021, Yang et al. 2021). In SERS, a nanostructured material is typically used as a substrate onto which the sample to be analyzed is placed. The substrate is typically fabricated by depositing nanoparticles (such as metallic nanospheres, nanocrystals, or nanofibers) on to a smooth plate to create a surface that is rough at the nanoscale level. This kind of nano-structured sensor has been used to determine the levels of artificial colors, pesticides, and toxins in foods using SERS (Zheng and He 2014).

5.2.2 Nanomaterials in electrochemical sensors

Magnetic nanomaterials (such as iron oxide nanoparticles) have great potential for fabricating electrochemical sensors for the analysis of various kinds of food components (Nejad et al. 2021). These devices can measure the concentration of specific analytes in a sample by using chemical or biochemical sensors that generate measurable electrical signals when a particular type of analyte interacts with the detection element on the surface of the sensor. Nanosized magnetic particles are often included at the surfaces of the electrodes due to their high specific surface areas, which increases the interaction between the detecting groups and the analytes, thereby improving the sensitivity of the sensor. Electrochemical devices typically contain a series of electrodes that can be used to monitor redox reactions involving the analyte of interest, which is usually dissolved in a solution. In some cases, biosensors are used (such as enzymes, nucleic acids, or antibodies) to detect specific kinds of molecules in a test sample. Electrochemical devices have been used to detect a broad range of substances in foods, including antioxidants in coffee, tea, wine, fruit, and vegetables, synthetic colorants in foods and beverages, pesticides in water, fruits, and vegetables, heavy metal ions in water and juices, and antibiotics in dairy, egg, meat, and seafood products (Nejad et al. 2021). The magnetic nanomaterials used to construct these sensors can be fabricated using several methods, including precipitation, thermal, microemulsion, ultrasonic, sol–gel, and biological techniques (Nejad et al. 2021). Iron oxides are one of the most common substances used for the fabrication of magnetic nanomaterials in electrochemical detectors.

5.2.3 Nanomaterials in colorimetric and fluorometric sensors

Nanomaterials can also be utilized to create sensors that undergo a color change when the detecting groups interact with a specific analyte (Li et al. 2019b). Gold nanoparticles are commonly used in this kind of colorimetric sensor (Liu et al. 2018, Li et al. 2019b). The color of a suspension of gold nanoparticles depends on their size, shape, surface properties, and aggregation state. For instance, a dispersion of small gold nanoparticles (10–50 nm) changes from red to blue when they go from a non-aggregated to an aggregated state (and *vice versa*). If the binding of a specific analyte to the surfaces of the nanoparticles leads to a change in their aggregation state, then there will be a corresponding change in color, which can be detected by the human eye. Typically, a specific receptor molecule is covalently attached to the surfaces of the gold nanoparticles. When a ligand binds to this receptor molecule, it causes a change in surface properties or aggregation state of the gold nanoparticles, thereby leading to a detectable color change. Fluorescent nanoparticles, such as quantum dots or carbon dots, can also be used as indicators of the binding of a specific type of molecule to receptor molecules attached to their surfaces. These kinds of colorimetric and fluorometric sensors have been employed to detect a wide range of analytes, including pathogens, pesticides, heavy metals, antibiotics, and mycotoxins in food and agricultural samples (Bulbul et al. 2015, Li et al. 2019b).

5.3 Nano-enabled packaging materials

The quality, freshness, safety, and shelf life of many foods are influenced by contamination, microbial growth, chemical reactions, and respiration events that occur during their production, storage, and distribution. Foods can be protected from contamination and degradation by using well-designed packaging materials (Khezerlou et al. 2021). At present, petroleum-based plastics are most commonly employed to fabricate the packaging materials used in the food industry due to their relatively low costs, ability for mass production, and good functional attributes (Alizadeh-Sani et al. 2019). However, the large-scale production and disposal of plastics is problematic due to their negative impacts on the environment and human health (Cazón et al. 2017). Consequently, there is interest in the creation of innovative packaging materials that can be used in the food industry that are more sustainable and environmentally friendly (Dehghani et al. 2018). Packaging materials assembled from biopolymers, such as polysaccharides or proteins, are being investigated as potential alternatives to conventional plastics (Dammak et al. 2017). Despite their better sustainability, biopolymer-based packaging materials do have some challenges to their widespread application. Their functional attributes, such as their optical, mechanical, barrier, or protective properties, are often insufficient for the intended purpose. The performance of these packaging materials can be enhanced by including additives in the biopolymer matrix, including crosslinkers, plasticizers, light blockers, diffusion blockers, antioxidants, antimicrobials, and sensors. These additives may be designed to remain within the packaging material or to slowly move into the food over time to provide a sustained effect, such as for preservatives (Ranjbaryan et al. 2019). Nevertheless, the incorporation of many types of additives into biopolymer-based films can be difficult due to their high volatility, poor water solubility, low chemical stability, poor matrix compatibility, or adverse effects on film properties (Ranjbaryan et al. 2019). Nanotechnology can often be used to overcome these challenges.

Some of the additives incorporated into biopolymer matrices to modulate their properties are nanoparticles themselves, such as the nanofibers or nanoparticles used to improve the mechanical or barrier properties of the films. Other additives may be converted into a nanoparticle form before being incorporated into the biopolymer matrices so as to improve their efficacy, such as hydrophobic colors, preservatives, or sensors (Esfanjani et al. 2018, Nejatian et al. 2019). Converting these substances into nanoparticles (or encapsulating them within nanoparticles) can reduce their volatility, enhance their chemical stability, increase their matrix compatibility, alter their transport properties, modulate their optical effects, and enhance their functional performance.

In this section, the various types of nanocarriers that can be used to encapsulate film additives is reviewed. Their effects on the physicochemical attributes of biopolymer-based packaging materials are then reviewed. Finally, several applications of nanotechnology to improve the functional performance of packaging materials are given.

5.3.1 Nanoencapsulation of film additives

A broad spectrum of nanoparticles have been investigated as potential carriers for film additives, such as microemulsions, nanoemulsions, nanoliposomes, solid lipid nanoparticles (SLN), nanostructured lipid carriers (NLC), biopolymer nanoparticles, and nanogels (Figure 5.1), which have been comprehensively reviewed elsewhere (McClements 2015, McClements and Ozturk 2021). For food packaging applications, these nanoparticles should be fabricated from materials that are compatible with foods and the environment. Ideally, they should have a low toxicity and should not cause damage to the environment after disposal of the packaging materials. For this reason, organic nanoparticles assembled from lipids, phospholipids, proteins, and/or polysaccharides are often used to encapsulate film additives (Sarkar et al. 2017, Maqsoudlou et al. 2020). Several kinds of fabrication methods have been developed and optimized for fabricating nanoparticles suitable for this purpose. These can be classified as top-down, bottom-up, or combination approaches (McClements 2015, Rezaei et al. 2019). Top-down approaches use physical, chemical, or enzymatic methods to break down larger substances into smaller nano-sized ones. Examples of this approach are high-pressure homogenizers or microfluidizers that convert bulk oil phases into nanosized oil droplets or milling devices that convert bulk

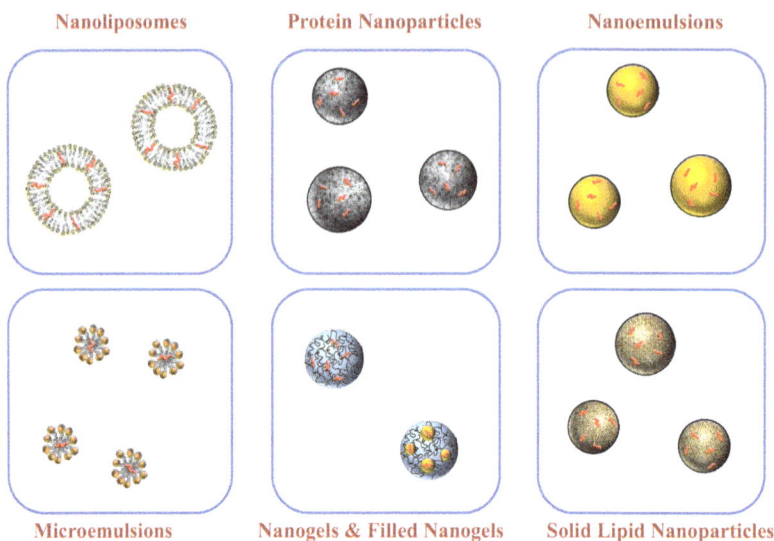

Figure 5.1: Various kinds of food-grade nanoparticles can be used to encapsulate bioactive components so that they can be used as functional additives in biopolymer-based films and coatings.

solid materials into solid nanoparticles. In contrast, bottom-up approaches use physiochemical methods to promote the assembly of molecules into nanosized particles. Examples of this approach include antisolvent precipitation to form biopolymer nanoparticles and spontaneous self-assembly of surfactant molecules to form microemulsions. Combination approaches utilize both top-down and bottom-up approaches. An example of this approach is the formation of nanolaminated oil droplets: small oil droplets are first produced by homogenization and then one or more layers of biopolymers are deposited on their surfaces using electrostatic deposition methods. The type of nanoparticle used to encapsulate an additive dictates the type of production method required to produce it.

Depending on the ingredients and processing methods used, the particle characteristics of nanocarriers can be controlled to meet different functional requirements (McClements 2015). For instance, their composition, physical state, dimensions, shape, electrical charge, surface chemistry, interfacial thickness, digestibility, and biodegradability can all be modulated (Figure 2.1). This provides manufacturers of packaging materials considerable scope to tailor the functional attributes of nanocarriers according to the nature of the additive that needs to be incorporated into the biopolymer matrix. For instance, the nanocarrier properties can be controlled to encapsulate additives with different polarities, volatilities, chemical stabilities, or functionalities, so they can be more easily introduced into packaging materials with different physicochemical properties, such as polarities, pore sizes, charges, mechanical rigidities, optical properties, and barrier properties (Figure 5.2). Several types of nanocarriers that can be used to incorporate functional additives into biopolymer-based packaging materials are highlighted here (Figure 5.1):

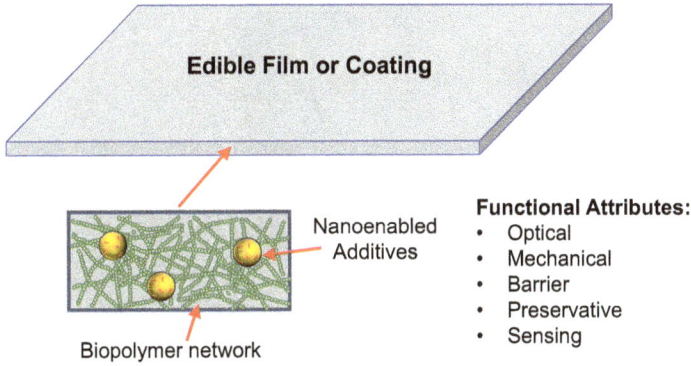

Edible Film or Coating

Nanoenabled Additives

Biopolymer network

Functional Attributes:
- Optical
- Mechanical
- Barrier
- Preservative
- Sensing

Figure 5.2: Biopolymer-based packaging materials (films and coatings) can be created from natural polymers (proteins and polysaccharides) and their functional performance can be enhanced by adding nanoenabled additives.

Biopolymer nanocarriers: This type of nanoparticle can be fabricated from polysaccharides (like alginate, carrageenan, cellulose, pectin, and starch) and/or proteins (like gelatin, casein, whey protein, soy protein, zein, and pea protein) (Chen et al. 2006, Fathi et al. 2014). These kinds of nanocarriers are sometimes divided into biopolymer nanoparticles and biopolymer nanogels. The main difference between them is that the former contains a relatively dense network of crosslinked biopolymer chains with little water, whereas the latter contains a relatively open network with a considerable amount of water trapped inside (Shewan and Stokes 2013). Hydrophilic bioactive agents can sometimes be encapsulated within biopolymer nanocarriers by trapping them within the particle interior through steric hindrance or physical attraction. For instance, the size of the bioactive agents may be greater than the pore size of the biopolymer network or there may be an attractive interaction between the bioactive agents and the biopolymer network. Alternatively, hydrophobic bioactive agents can be encapsulated within lipid-based nanoparticles first (such as micelles or nanoemulsions) and then incorporated into biopolymer nanoparticles or microparticles. The retention and release of active agents from biopolymer nanocarriers can be controlled in a variety of fashions, such as simple diffusion, swelling, disintegration, or altering the molecular interactions with the biopolymer network (Figure 5.3).

Nanoliposomes: This type of nanoparticle is usually fabricated from phospholipids, which may be isolated from animal (*e.g.*, milk or egg) or plant (*e.g.*, soy or sunflower) sources (Fathi et al. 2012). Structurally, these nanoparticles contain a single or several phospholipid bilayers arranged into onion-like concentric shells that

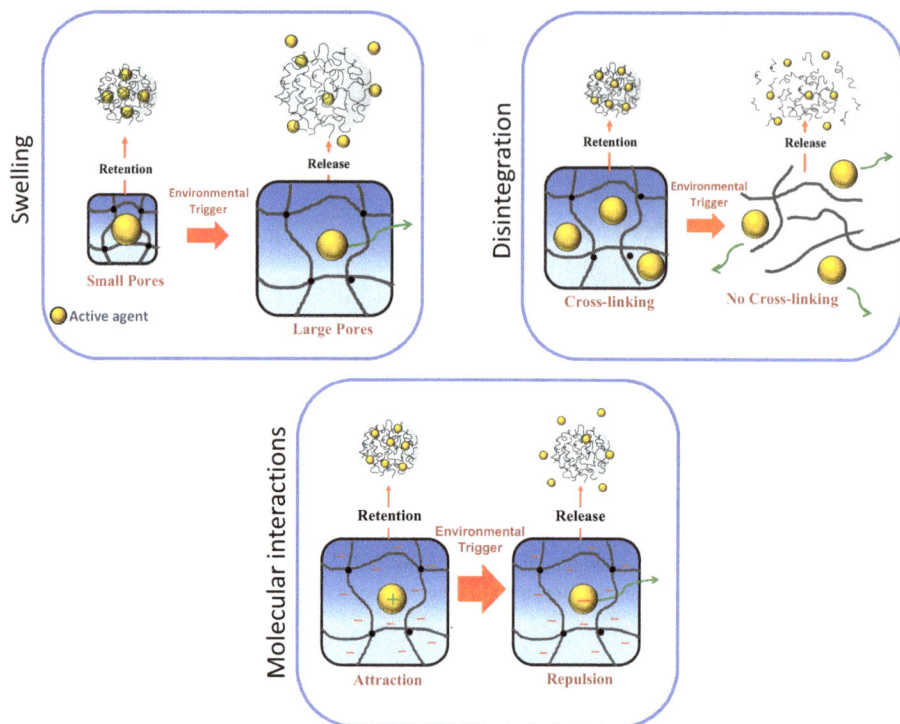

Figure 5.3: Biopolymer nanocarriers can be designed to retain and release active agents through different mechanisms, including swelling, disintegration, and molecular interactions.

surround a watery core (Figure 5.1). Consequently, these kinds of nanoparticles have both hydrophobic domains (phospholipid tails) and hydrophilic domains (phospholipid heads and water), which means they can encapsulate non-polar and polar bioactive molecules. The functionality of nanoliposomes can be modulated by using phospholipids with different head or tail groups. For instance, hydrocarbon tail groups with different chain lengths or degrees of unsaturation have different melting points. Consequently, the rigidity and barrier properties of the phospholipid bilayers can be controlled by using different types of phospholipids. They may be useful for controlling the retention and release of encapsulated active agents in response to temperature changes. A substance may be retained at temperatures below the thermal transition temperature of the bilayer but be released at higher temperatures. The properties of nanoliposomes can also be modulated by coating them with one or more biopolymer layers using electrostatic attraction.

Nanoemulsions: This type of lipid nanoparticle is often used to incorporate lipophilic bioactive agents within packaging materials, especially essential oils. They contain small emulsifier-coated liquid oil droplets that are dispersed within water (Figure 5.1). Thus, they are typically constructed from oil, water, and emulsifiers

(McClements and Rao 2011, Silva et al. 2012, Rezaei et al. 2019). Typically, they contain oil droplets that have dimensions ranging from around 50 nm to 500 nm depending on the homogenization method and ingredients used to formulate them, but smaller ones can be produced under some circumstances. Nanoemulsions containing sufficiently small droplets (<50 nm) tend to be optically clear, whereas those containing larger droplets tend to be turbid or opaque, which may be important for controlling the appearance and light blocking properties of packaging materials.

Lipid nanosuspensions: These types of suspensions contain emulsifier-coated lipid particles suspended in water (like nanoemulsions) but the interior of the particles is fully or partially solidified (Figure 5.1). The two most important kinds of lipid nanosuspensions are SLN, which contain fully crystalline particle interiors, and NLCs, which contain partially crystalline ones (Weiss et al. 2008, Tamjidi et al. 2013). Hydrophobic bioactive agents are typically introduced into the lipid phase prior to the formation of the SLNs or NLCs. The fact that the lipid phase is solidified can slow down the chemical degradation of encapsulated substances as well as increasing their retention and slowing or controlling their release profiles. After preparation, bioactive-loaded SLNs or NLCs can be introduced into food packaging materials to enhance their functional properties (Brandelli et al. 2017).

The selection of a particular kind of nanoparticle to introduce substances into packaging materials is governed by several considerations, such as their loading capacities, encapsulation efficiencies, stabilization attributes, light blocking properties, aggregation state, and matrix compatibility. Different kinds of nanoparticles have specific advantages and disadvantages for specific applications.

5.3.2 Nano-enabled additives used in packaging materials

Active and smart packaging materials are being designed to have new or improved functional attributes compared to conventional packaging materials (Ahari and Soufiani 2021, Sani et al. 2021a, Siddiqui et al. 2022). For instance, they can be designed to contain preservatives (such as antimicrobials or antioxidants) to extend the shelf life of foods, light blockers (scatterers or absorbers) to protect foods against photodegradation, or sensors (such as pH-, light-, or temperature-responsive devices) to provide information about the quality, safety, or freshness of packaged foods. In this section, the focus is on the use of nanotechnology to incorporate functional components into food packaging materials, so as to improve their functional attributes.

5.3.2.1 Antimicrobials

Synthetic or natural antimicrobial agents can be incorporated into packaging materials to inhibit the growth of spoilage or pathogenic microorganisms, such as bacteria, yeasts, or molds (Matthews et al. 2017). However, there is a growing interest in the utilization of natural (especially plant-based) antimicrobials due to consumer demand for a more sustainable, healthy, and ethical food supply (Gutierrez-del-Rio et al. 2018). Many plants produce and secrete secondary metabolites with antimicrobial activity to protect themselves against microbial contamination (Jafarzadeh et al. 2020, Wink 2022). These phytochemicals are often essential oils or other phytochemicals, like polyphenols (Gutierrez-del-Rio et al. 2018, Yadav et al. 2020). Many of these substances are challenging to introduce into biopolymer-based packaging materials due their high volatility, intense odors, poor chemical stability, and/or poor water-solubility. For these reasons, they often need to be encapsulated within nanoenabled delivery systems (such as nanoemulsions, nanoliposomes, or lipid nanosuspensions) before they can be introduced into packaging materials (Li et al. 2019a). Essential oils have been widely studied as natural plant-based preservatives for use in food and packaging applications (Rao et al. 2019). They have been demonstrated to show strong antimicrobial activity against a broad spectrum of spoilage and pathogenic microbes. These effects have been attributed to a variety of physicochemical and physiological mechanisms, including disruption of the phospholipid bilayers in microbial cell membranes, disturbance of key proteins embedded in these bilayers (such as receptor, signaling, or transport molecules), and interaction with critical enzymes in biochemical pathways within the cells (Hassoun et al. 2020). For instance, cinnamaldehyde, eugenol, and thymol have all been incorporated into packaging materials to inhibit microbial growth during storage (Tonyali et al. 2020). The phytochemical curcumin has also been shown to have potential applications as an antimicrobial agent in biodegradable packaging materials due to its ability to inhibit the growth of a variety of microorganisms (Papadimitriou et al. 2018).

To be effective, an antimicrobial agent must be present at a sufficiently high initial concentration within the packaging material, it should be compatible with the polymer matrix, it must remain chemically stable during storage, it should not be lost due to volatilization, and it should be able to move to the place where the microorganisms are located. These challenges can often be overcome using well-designed nanoenabled delivery systems. Further information about the kinds of phytochemicals suitable as antimicrobial additives for food packaging applications can be found elsewhere (Seow et al. 2014, Calo et al. 2015, Donsi and Ferrari 2016, Gutierrez-del-Rio et al. 2018).

5.3.2.2 Antioxidants

Some foods contain appreciable levels of unsaturated fats or proteins that are prone to chemical degradation due to oxidative reactions, which decreases their quality, shelf life, and nutritional profile (Sun and Holley 2012). These oxidative reactions can often be inhibited by incorporating synthetic or natural antioxidants into packaging materials (Valdés et al. 2015). As with antimicrobials, there is a growing trend within the food industry to use natural (particularly plant-based) antioxidants in food packaging. Many of the phytochemicals that are effective antimicrobials are also effective antioxidants, including essential oils and polyphenols. Essential oils isolated from various botanical sources have been demonstrated to have good antioxidant activity when integrated into packaging materials (Valdés et al. 2015). For instance, researchers have shown that oregano oil exhibits strong antioxidant properties when incorporated into packaging materials, which can retard food deterioration during storage (Al-Bandak and Oreopoulou 2007). Polyphenols such as quercetin have also been shown to have good antioxidant properties in packaging materials (Tavassoli et al. 2021). Many natural antioxidants are strongly hydrophobic molecules that cannot easily be incorporated into polymer films. Consequently, they need to be encapsulated within suitable nanoparticles before they can be introduced into the films. Several types of natural antioxidants have been incorporated into different kinds of nanoparticles for this reason, such as rutin-loaded lipid nanodroplets in gelatin films (Dammak et al. 2017), vitamin E-loaded lipid nanodroplets in whey protein films (Agudelo-Cuartas et al. 2020), and clove oil-loaded lipid nanodroplets in pullulan-gelatin films (Shen et al. 2021). Incorporation of these antioxidant nanoparticles into the films has largely been linked to the fact that essential oils and phytochemicals contain numerous hydroxyl groups that act as electron donors to free radicals, which converts them into less reactive species, thereby breaking the free radical chain reaction and slowing down oxidation (Contreras-Calderón et al. 2011, Darbasi et al. 2017). More information about the different types of essential oils and other phytochemicals that can be utilized as natural antioxidants in packaging applications can be found in several comprehensive review articles (Cardoso-Ugarte and Sosa-Morales, Amorati and Valgimigli 2018, Ferrentino et al. 2020, Basavegowda and Baek 2021, Gutierrez-del-Rio et al. 2021).

5.3.2.3 Light blockers

Some foods contain ingredients that are susceptible to photodegradation, especially in the presence of ultraviolet light. Consequently, it is useful to incorporate substances that can block this form of electromagnetic radiation from getting into packaging materials (Figure 5.4) (Sani et al. 2021a). Light blockers can do this in two ways depending on their physicochemical properties: (i) particulate substances can scatter light waves; (ii) chromophores can absorb light waves. Normally, particles and chromophores scatter or absorb light over a particular range of wavelengths depending on their characteristics. For particles, the scattering *versus* wavelength profile depends on their size relative to the wavelength of light, whereas for chromophores, the absorption *versus* wavelength profile depends on the nature of the chemical groups them contain (especially unsaturated ones). There is therefore interest in introducing light blockers into food packaging materials to improve the shelf life and quality of foods containing photodegradable substances (Stoll et al.

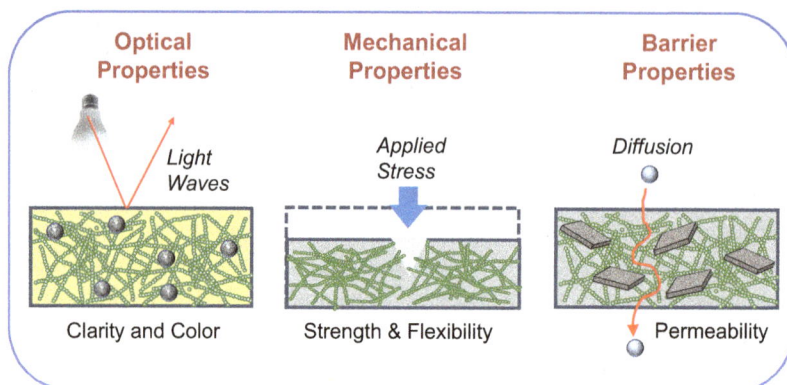

Figure 5.4: Biopolymer-based packaging materials must be able to match the optical, mechanical, and barrier properties of conventional plastics, which can often be achieved by including nano-enabled additives.

2021), especially polyunsaturated lipids that are susceptible to oxidative damage (Pristouri et al. 2010). Carotenoids, such as bixin, are natural pigments that can be incorporated into packaging materials as light blockers due to their ability to absorb light waves (Scotter 2009). However, carotenoids are strongly hydrophobic molecules that have a very low solubility in water. Consequently, it is often necessary to incorporate them into lipid nanoparticles before they can be introduced into biopolymer-based films and coatings, which tend to be predominantly hydrophilic. The same is true for other kinds of hydrophobic pigments, such as curcumin. Nanoparticles themselves can also be used to reduce the amount of light waves that can pass through packaging materials due to their ability to strongly scatter electromagnetic radiation. However, the size of the nanoparticles must be carefully controlled so that they scatter light over the range of wavelengths that are most likely to promote photodegradation. In principle, it is possible to use a mixture of nanoparticles and chromophores to both scatter and absorb light. However, it is important that these light blocking substances do not adversely impact other important functional attributes of the packaging materials.

5.3.2.4 Film strengtheners

The mechanical attributes of packaging materials, including their tensile strength, elongation at break, and elastic modulus, also influence their performance (Alizadeh-Sani et al. 2018). For instance, they play an important role in protecting foods from physical damage throughout storage and transport. However, they should also be able to be easily handled or removed by the consumer. In general, the mechanical attributes of packaging materials depend on the kinds of substances (usually polymers) used to construct them as well as the type, number, and magnitude of the interactions acting amongst these substances (Sanchez-Gonzalez et al. 2011, Chen et al. 2016). Biopolymer-based packaging materials, such as those constructed from proteins and/or polysaccharides, often do not have the required mechanical attributes for food or non-food applications (Azeredo and Waldron 2016). For instance, they may be too soft or too brittle. For this reason, there has been interest in increasing the strength and flexibility of these kinds of packaging materials by incorporating nanomaterials within them. These nanomaterials may be nanofibers, nanoparticles, or nanosheets (Figure 5.5). For instance, the mechanical properties of films made from soy proteins have been improved by incorporating cellulose nanofibers within them (Wei et al. 2021). Similarly, the mechanical properties of whey protein films have been enhanced by incorporating nanocellulose crystals in them (Qazanfarzadeh and Kadivar 2016). Nanocellulose has also been used to improve the mechanical properties of starch-based films (Bangar and Whiteside 2021). In some cases, combinations of nanoparticles have been used to improve both the mechanical and barrier properties of films. For instance, a combination of nanocellulose crystals and zinc oxide nanoparticles has been used to strengthen soy protein films as well as reduce their permeability (Xiao et al. 2020). In general, the type and concentration of nanomaterials used must be optimized to obtain films with the required physicochemical and functional attributes. For example, it has been shown that increasing the concentration of nanomaterials in biopolymer films first strengthens but then weakens them, and so the concentration employed must be optimized (Bangar and Whiteside 2021).

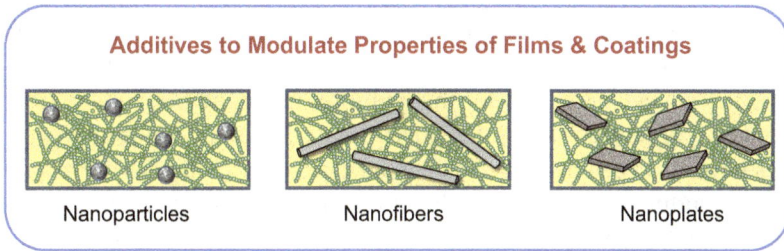

Additives to Modulate Properties of Films & Coatings

Nanoparticles   Nanofibers   Nanoplates

Figure 5.5: The optical, mechanical, and barrier properties of biopolymer-based packaging materials, such as their appearance, strength, flexibility, and permeability, can be modulated by adding different kinds of nanoenabled additives, including particles, fibers, and plates.

In some applications, the ability of films to resist changes in their properties when the temperature is changed is important. Studies have shown that the thermal behavior of biopolymer-based films can often be improved by introducing nanomaterials into them. For instance, the phase transition temperatures and thermal stability of biopolymer films can be modulated by adding nanomaterials. Incorporating lipid nanoparticles into carboxymethyl cellulose films was shown to reduce their melting temperature (Mirzaei-Mohkam et al. 2020). Introducing nanoliposomes into pullulan films was shown to decrease their glass transition temperature (Nejatian et al. 2019). In another study, it was shown that incorporating SLN into polymer films increased their thermal stability (de Carvalho et al. 2019). In general, these studies indicate that the thermal responses of biopolymer-based films can be controlled by incorporating different kinds of nanoparticles into them. However, further research is still required to identify the physicochemical origin of these effects.

5.3.2.5 Gas barriers

The diffusion of gasses through packaging materials, especially oxygen and moisture, influences the shelf life and quality of many foods (Figure 5.4). Controlling the diffusion of oxygen through packaging materials and into foods is often desirable because it can promote oxidative reactions (Sahraee et al. 2019). Controlling the diffusion of water vapor is also important because it may impact the water activity of packaged foods, thereby influencing their texture, microbial growth, and chemical reactions. The gas permeability of biopolymer-based films is often too high for many commercial applications. The gas barrier properties of these films can often be improved by incorporating inorganic and/or organic nanomaterials within them, such as TiO_2-, ZnO-, silver-, cellulose-, chitin-, starch-, protein-, or lipid-based nanomaterials (Sun et al. 2018, Helanto et al. 2019, Jafarzadeh and Jafari 2021, Dash et al. 2022). These nanomaterials may come in different sizes, shapes, and surface chemistries, which alters their barrier properties, as well as their ability to be successfully incorporated into the biopolymer matrix. Nanomaterials can occupy the pores within the 3D network formed by the biopolymer molecules or they can simply act as obstructions, which inhibits the diffusion of molecules through the films by increasing the pathlength they must travel (Figure 5.4). Various kinds of inorganic nanoparticles have been used for this purpose, including those made from TiO_2, ZnO, and SiO_2 (Jafarzadeh and Jafari 2021). Similarly, several kinds of organic nanoparticles have also been used to enhance the gas barrier properties of packaging materials, including lipid nanoparticles (Dammak et al. 2017), nanocellulose (Chowdhury et al. 2019), and nanochitin (Ghorbani et al. 2015, Hosseini et al. 2016, Sani et al. 2021b).

5.3.2.6 Sensors/indicators

Smart packaging materials are being created by incorporating substances that can provide an indication of the quality, freshness, or safety of food products into biopolymer-based matrices to provide food distributors and consumers real-time information about food properties (Figure 5.6) (Grillo et al. 2021). These substances are often natural pigments, like carotenoids or anthocyanins, which can change color in response to alterations in their environments (Mustafa and Andreescu 2018). For instance, alterations in the temperature, pH, or gas levels within a food or package can promote a change in the color of the pigments, which can be used as an indication of

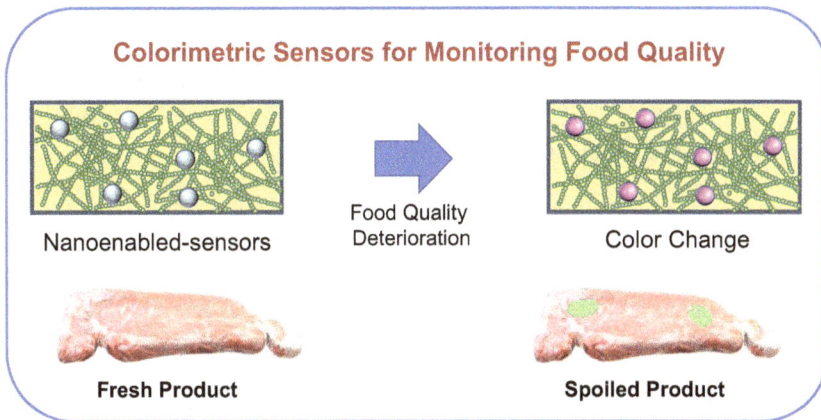

Figure 5.6: Changes in the quality, freshness, or safety of packaged food products can sometimes be monitored in real time using nanostructured packaging materials containing sensors that change color in response to changes in food properties such as freshness or safety.

food properties. In some cases, these natural pigments need to be encapsulated within suitable nanocarriers to increase their compatibility with the biopolymer matrix, to enhance their chemical stability, or to improve their performance.

5.3.2.7 Water barrier enhancers

The interaction of packaging materials with moisture in their environment, which could be from the food itself or from its surroundings, is also an important attribute (Fathi et al. 2018). For instance, some films stay intact when they contact water, whereas others absorb moisture and dissolve. Biopolymer-based films are particularly susceptible to absorbing water and dissolving, which often limits their application as replacements for petroleum-based films (Pirouzifard et al. 2020). This problem can be overcome by incorporating nanomaterials into the film matrix, such as nanoparticles, nanofibers, or nanoplates (Figure 5.5), especially the ones that are impermeable to water (Pérez-Córdoba et al. 2018). It should be noted that there are some applications where the tendency for films to dissolve in water is an advantage, *e.g.*, for the development of films that melt in the mouth to release flavors, or films that release preservatives (such as antimicrobials or antioxidants) when the food comes into contact with water (Khezerlou et al. 2019). Researchers have reported that incorporating various kinds of nanoparticles into biopolymer-based films can reduce their water solubility, including lipid nanoparticles (Li et al. 2020, Behjati and Yazdanpanah 2021), nanocellulose (Ahmad et al., Kocabas et al. 2021), zinc oxide nanoparticles (Yadav et al. 2021), and titanium dioxide (Erciyes and Ocak 2019).

5.3.3 Application of nano-enabled packaging materials into foods

The potential application of nano-enabled packaging materials to enhance the quality, shelf-life, and safety of foods is highlighted in this section by giving several examples from different food categories.

5.3.3.1 Seafood

Seafood products, like fish and shellfish, are highly perishable due to the fact that they have a high moisture content and contain macronutrients and micronutrients that microorganisms can utilize to grow and survive (Tsironi and Taoukis 2018, Sani et al. 2021a). In addition, many seafood products contain high levels of polyunsaturated lipids (like omega-3 fats) that can degrade due to oxidation reactions, thereby leading to rancidity. Consequently, it is important to use packaging materials that can inhibit and/or detect their degradation.

Microbial activity and lipid oxidation generate a range of characteristic reaction products that can be used as markers of seafood quality during storage (Tsironi and Taoukis 2018). For instance, microbial degradation leads to the production of dimethylamine, trimethylamine, and ammonia, whereas oxidation leads to the formation of hydroperoxides, conjugated dienes, and aldehydes that can be determined using suitable analytical techniques. Researchers have shown that nano-enabled smart packaging materials can be designed to provide information about seafood quality by incorporating sensors into them that give a measurable change in response to the presence of one or more of these substances. For instance, natural pigments have been incorporated into these packaging materials that change color in response to changes in pH or ammonia gas production, thereby providing an indication of product freshness. As an example, researchers developed smart packaging materials by incorporating curcumin-loaded nanoparticles into biopolymer films that could detect the freshness of shrimp during storage (Xiao et al. 2020). These films changed from bright yellow to reddish brown as the shrimp deteriorated.

Nanotechnology has also been used to incorporate natural antioxidants and/or antimicrobials into active packaging materials to increase the shelf life of seafood products by retarding oxidative deterioration or by inhibiting microbial growth. Eugenol oil nanoemulsions applied to the surfaces of shrimp as part of edible coatings have been shown to inhibit the rate of lipid oxidation (Sharifimehr et al. 2019) and microbial growth (Nazari et al. 2019). Similarly, the shelf-life of fish fillets during cold storage has been increased by incorporating cumin oil nanoliposomes into nanostructured edible coatings (Homayonpour et al. 2021).

5.3.3.2 Meat

Fresh meat is also a high moisture food containing a diversity of nutrients, thereby making it prone to deterioration through microbial growth and oxidative reactions

(Lambert et al. 1991, Sani et al. 2021a). Consequently, there is also interest in the development of smart and/or active packaging materials to monitor and control these degradative processes in packaged meat products. The inclusion of various kinds of essential oil nanoemulsions into biopolymer-based edible coatings and films has been demonstrated to improve the quality and increase the shelf life of turkey (Kazemeini et al. 2021), chicken (Kamkar et al. 2021), beef (Xavier et al. 2021), and pork (Xiong et al. 2020). These effects are mainly due to the antioxidant and antimicrobial activities of the phytochemicals within the essential oils. Other kinds of nanoparticles incorporated into biodegradable packaging materials have also been shown to have preservative properties, including nanochitosan and zinc oxide nanoparticles (Amjadi et al. 2020). Smart packaging materials have also been developed to monitor changes in the freshness of meat during storage by incorporating pH-responsive color sensors (such as anthocyanins) into nano-enabled biopolymer-based films (Taherkhani et al. 2020, Zhai et al. 2020, Alizadeh-Sani et al. 2021). These sensors give a color change when the meat deteriorates due to alterations in the pH of the product as well as due to gas production by the muscle tissues as they chemically degrade.

5.3.3.3 Cheese

Cheese is rich in proteins and lipids, making it prone to contamination by microbes such as bacteria, yeast, and mold (Costa et al. 2018). Several researchers have shown that nanotechnology can be used to increase the shelf-life of cheese products by incorporating antimicrobial nanoparticles into packaging or coating materials (Lima et al. 2021). As an example, thyme-oil loaded nanoliposomes have been introduced into chitosan-based coatings to increase the shelf life of cheese (Al-Moghazy et al. 2021). These coatings did not impact the sensory quality of the cheese, but they did extend the time that the product retained its desirable quality attributes by inhibiting microbial growth. Zinc oxide and clay nanoparticles have also been incorporated into biopolymer-based films to increase the shelf life of cheese by inhibiting microbial growth (Jafarzadeh et al. 2019). A combination of silver nanoparticles and lemongrass oil have also been shown to provide strong antimicrobial effects when incorporated into biopolymer-based packaging materials used to protect cheese (Motelica et al. 2021).

5.3.3.4 Bread

The quality and shelf life of baked products like bread can also be improved by developing innovative nano-enabled packaging materials. For instance, biopolymer-films loaded with carvacrol or oregano oil nanoemulsions were shown to increase the shelf stability of bread stored under ambient conditions (Otoni et al. 2014, Lei et al. 2019). This effect was mainly attributed to the ability of the essential oils to inhibit oxidation and microbial growth due to their antioxidant and antimicrobial activities. In particular, these natural substances are effective at retarding the growth of yeasts and molds, which often cause spoilage of bread. Inorganic nanoparticles have also been shown to be effective preservatives for application in active packaging materials. For instance, the incorporation of zinc oxide nanoparticles into biopolymer-based films has been shown to increase the shelf life of bread by inhibiting microbial growth (Noshirvani et al. 2017). Similarly, silver nanoparticles incorporated into biopolymer-based based films were shown to extend the shelf life of bread (Nair et al. 2017). These inorganic nanoparticles were shown to be effective against the growth of both yeasts and molds.

5.3.3.5 Fruit *and vegetables*

Fruits and vegetables may also suffer quality losses throughout storage due to physical, chemical, or biological changes, including transpiration, respiration, ethylene production, and fungal growth (Duan et al. 2019). Ideally, these processes could be controlled by using biopolymer-based packaging materials, but these materials often do not exhibit appropriate functional attributes. Consequently, there has been interest in using nanotechnology to improve the performance of biopolymer-based films and coatings. The quality and shelf life of cut melon has been improved by coating them with a solution containing β-carotene nanoparticles and xanthan gum (Zambrano-Zaragoza et al. 2017). Thyme oil-loaded zein nanofibers have been used to retard fungal and yeast growth on fresh strawberries, which extended their shelf life (Ansarifar and Moradinezhad 2021). Many other kinds of essential oils have also been loaded into nanoparticles and then incorporated into packaging materials for the same purpose (Niu et al. 2018, Al-Tayyar et al. 2020, Teixeira et al. 2022). Inorganic antimicrobial nanoparticles have also been incorporated into packaging materials to improve the quality and shelf life of fruits and vegetables, including those composed of silver (Basumatary et al. 2018), zinc oxide, copper oxide (Kalia et al. 2021), and silicon dioxide (Sami et al. 2021). However, from both consumer perception and environmental viewpoints, it would be better to utilize organic nanoparticles for this purpose.

Overall, these results show that nanotechnology has great potential to improve the functional performance of biopolymer-based films and coatings, which may lead to a new generation of more sustainable and environmentally friendly packaging materials to replace plastics. Nevertheless, further research and development is still required to create robust packaging materials that can be produced economically on the scale required for commercial applications.

5.4 Nano-enabled filters

As discussed in Chapter 3, certain kinds of nanomaterials are particularly useful for creating filtering devices because of their high specific surface areas and tunable pore sizes and surface chemistries (Kumar et al. 2019, Badgar et al. 2022). As a result, they can be used to adsorb relatively large quantities of materials onto their surfaces, or they can be used to discriminate between molecules with different dimensions or polarities by controlling the pore size and surface chemistry of the nanofiber network. Organic nanofibers, such as nanocellulose and nanochitin, have been widely explored for this purpose (Douglass et al. 2018, Naghizadeh and Ghofouri 2019, Mautner 2020, Sharma et al. 2020, Liu et al. 2021). These materials are obtained by chemical and/or enzymatic modification of natural resources such as wood, cotton, bacteria, and agricultural waste in the case of nanocellulose, and from crustacean shells in the case of nanochitin. Nanofilters can be utilized to remove pollutants from wastewater, or they can be used to extract compounds that can be converted into valuable ingredients (such as flavors, dietary fibers, proteins, vitamins, or nutraceuticals) from waste streams (Kotsanopoulos and Arvanitoyannis 2015, Castro-Munoz et al. 2020). Consequently, they have a broad range of potential applications within the food and agricultural industry as well as other industries.

5.5 Miscellaneous applications

Advanced nanomaterials created from food waste materials have been developed for various other applications. For instance, bioplastics created from nanocellulose and nanochitin are being created to replace conventional petroleum-based plastics in products such as utensils, shoes, electronics, and consumer goods (Otoni et al. 2021). Organic nanomaterials derived from natural sources are also being used to create novel optical devices, such as solar cells, diodes, and sensors (Kaschuk et al. 2022). As discussed elsewhere, proteins, polysaccharides, and lipids are being utilized to create nanoparticles that can be used to encapsulate, protect, and deliver drugs in the pharmaceutical industry as well as to replace conventional synthetic polymers or surfactants in a broad range of other applications (Amiri et al. 2021, Gomes and Sobral 2022).

5.6 Conclusions and future directions

Nanotechnology is increasingly being used to create innovative materials, especially those made from biodegradable substances, such as food proteins, polysaccharides, phospholipids, and lipids. For instance, they have been used to create sensors with enhanced selectively or sensitivity by utilizing nanomaterials that have large specific surface areas, whose surface properties can be physically or chemically modified to make them responsive to specific substances or to changes in their environment. Another major application of nanotechnology has been in the development of smart or active packaging materials that can be used to provide real-time information about the quality, freshness, or safety of packaged foods, or that can improve these attributes by including preservatives within the films or coatings. Many of these innovative packaging materials are being developed from natural polymers, such as proteins and polysaccharides, which should make them more environmentally friendly and sustainable than those made from petroleum-based polymers. Novel nanomaterials assembled from agriculturally derived substances are also being developed for applications in the food, agriculture, and other industries. Nevertheless, further research and development are still required to ensure these products are economically feasible, mass producible, and meet the technical requirements needed for commercial applications. In the future, it is likely that nanotechnology will continue to be used to improve the functional performance of existing materials or to create new materials with novel properties and applications. However, it will be important to ensure that these nanomaterials are safe to produce and use, and that they do not cause any adverse effects on the environment, which is the subject of the following chapter. Moreover, it is critical to ensure that these nanomaterials have the robustness and functional performances required for commercial applications as well as being capable of large-scale and economic production.

References

Agudelo-Cuartas, C., D. Granda-Restrepo, P. J. A. Sobral, H. Hernandez and W. Castro (2020). "Characterization of whey protein-based films incorporated with natamycin and nanoemulsion of α-tocopherol". Heliyon 6(4): e03809.

Ahari, H. and S. P. Soufiani (2021). "Smart and active food packaging: insights in novel food packaging". Frontiers in Microbiology 12.

Ahmad, K., Z. U. Din, H. Ullah, Q. Ouyang, S. Rani, I. Jan, M. Alam, Z. Rahman, T. Kamal, S. Ali, S. A. Khan, D. Shahwar, F. Gul, M. Ibrahim and T. Nawaz "Preparation and characterization of bio-based nanocomposites packaging films reinforced with cellulose nanofibers from unripe banana peels". Starch-Starke.

Al-Bandak, G. and V. Oreopoulou (2007). "Antioxidant properties and composition of Majorana syriaca extracts". European Journal of Lipid Science and Technology 109(3): 247–255.

Al-Moghazy, M., H. S. El-sayed, H. H. Salama and A. A. Nada (2021). "Edible packaging coating of encapsulated thyme essential oil in liposomal chitosan emulsions to improve the shelf life of Karish cheese". Food Bioscience 43: 101230.

Al-Tayyar, N. A., A. M. Youssef and R. R. Al-Hindi (2020). "Edible coatings and antimicrobial nanoemulsions for enhancing shelf life and reducing foodborne pathogens of fruits and vegetables: A review". Sustainable Materials and Technologies 26.

Alizadeh-Sani, M., A. Ehsani, E. M. Kia and A. Khezerlou (2019). "Microbial gums: Introducing a novel functional component of edible coatings and packaging". Applied microbiology and biotechnology 103(17): 6853–6866.

Alizadeh-Sani, M., A. Khezerlou and A. Ehsani (2018). "Fabrication and characterization of the bionanocomposite film based on whey protein biopolymer loaded with TiO2 nanoparticles, cellulose nanofibers and rosemary essential oil". Industrial crops and products 124: 300–315.

Alizadeh-Sani, M., M. Tavassoli, E. Mohammadian, A. Ehsani, G. J. Khaniki, R. Priyadarshi and J. W. Rhim (2021). "pH-responsive color indicator films based on methylcellulose/chitosan nanofiber and barberry anthocyanins for real-time monitoring of meat freshness". International Journal of Biological Macromolecules **166**: 741–750.

Amiri, M. S., V. Mohammadzadeh, M. E. T. Yazdi, M. Barani, A. Rahdar and G. Z. Kyzas (2021). "Plant-based gums and mucilages applications in pharmacology and nanomedicine: a review". Molecules **26**(6): 1770, 1–17. doi.org/10.3390/molecules26061770.

Amjadi, S., M. Nazari, S. A. Alizadeh and H. Hamishehkar (2020). "Multifunctional betanin nanoliposomes-incorporated gelatin/chitosan nanofiber/ZnO nanoparticles nanocomposite film for fresh beef preservation". Meat science **167**: 108161.

Amorati, R. and L. Valgimigli (2018). "Methods to measure the antioxidant activity of phytochemicals and plant extracts". Journal of Agricultural and Food Chemistry **66**(13): 3324–3329.

Ansarifar, E. and F. Moradinezhad (2021). "Preservation of strawberry fruit quality via the use of active packaging with encapsulated thyme essential oil in zein nanofiber film". International Journal of Food Science & Technology **56**(9), 4239–4247.

Azeredo, H. M. C. and K. W. Waldron (2016). "Crosslinking in polysaccharide and protein films and coatings for food contact – A review". Trends in Food Science & Technology **52**: 109–122.

Badgar, K., N. Abdalla, H. El-Ramady and J. Prokisch (2022). "Sustainable applications of nanofibers in agriculture and water treatment: a review". Sustainability **14**(1): 464, 1–15. doi.org/10.3390/su14010464

Bangar, S. P. and W. S. Whiteside (2021). "Nano-cellulose reinforced starch bio composite films- A review on green composites". International Journal of Biological Macromolecules **185**: 849–860.

Basavegowda, N. and K. H. Baek (2021). "Synergistic Antioxidant and Antibacterial Advantages of Essential Oils for Food Packaging Applications". Biomolecules **11**(9): 1267, 1–12. https://doi.org/10.3390/biom11091267

Basumatary, K., P. Daimary, S. K. Das, M. Thapa, M. Singh, A. Mukherjee and S. Kumar (2018). "Lagerstroemia speciosa fruit-mediated synthesis of silver nanoparticles and its application as filler in agar based nanocomposite films for antimicrobial food packaging". Food Packaging and Shelf Life **17**: 99–106.

Behjati, J. and S. Yazdanpanah (2021). "Nanoemulsion and emulsion vitamin D3 fortified edible film based on quince seed gum". Carbohydrate Polymers **262**: 117948.

Brandelli, A., L. F. W. Brum and J. H. Z. Dos Santos (2017). "Nanostructured bioactive compounds for ecological food packaging". Environmental Chemistry Letters **15**(2): 193–204.

Bulbul, G., A. Hayat and S. Andreescu (2015). "Portable Nanoparticle-Based Sensors for Food Safety Assessment". Sensors **15**(12): 30736–30758.

Calo, J. R., P. G. Crandall, C. A. O'Bryan and S. C. Ricke (2015). "Essential oils as antimicrobials in food systems – A review". Food Control **54**: 111–119.

Cardoso-Ugarte, G. A. and M. E. Sosa-Morales "Essential Oils from Herbs and Spices as Natural Antioxidants: Diversity of Promising Food Applications in the past Decade". Food Reviews International, 1–17. doi.org/10.1080/87559129.2021.1872084

Castro-Munoz, R., G. Boczkaj, E. Gontarek, A. Cassano and V. Fila (2020). "Membrane technologies assisting plant-based and agro-food by-products processing: A comprehensive review". Trends in Food Science & Technology **95**: 219–232.

Cazón, P., G. Velazquez, J. A. Ramírez and M. Vázquez (2017). "Polysaccharide-based films and coatings for food packaging: A review". Food Hydrocolloids **68**: 136–148.

Chen, H., X. Hu, E. Chen, S. Wu, D. J. McClements, S. Liu, B. Li and Y. Li (2016). "Preparation, characterization, and properties of chitosan films with cinnamaldehyde nanoemulsions". Food Hydrocolloids **61**: 662–671.

Chen, L. Y., G. E. Remondetto and M. Subirade (2006). "Food protein-based materials as nutraceutical delivery systems". Trends in Food Science & Technology **17**(5): 272–283.

Chowdhury, R. A., M. Nuruddin, C. Clarkson, F. Montes, J. Howarter and J. P. Youngblood (2019). "Cellulose Nanocrystal (CNC) Coatings with Controlled Anisotropy as High-Performance Gas Barrier Films". Acs Applied Materials & Interfaces **11**(1): 1376–1383.

Contreras-Calderón, J., L. Calderón-Jaimes, E. Guerra-Hernández and B. García-Villanova (2011). "Antioxidant capacity, phenolic content and vitamin C in pulp, peel and seed from 24 exotic fruits from Colombia". Food research international **44**(7): 2047–2053.

Costa, M. J., L. C. Maciel, J. A. Teixeira, A. A. Vicente and M. A. Cerqueira (2018). "Use of edible films and coatings in cheese preservation: Opportunities and challenges". Food Research International **107**: 84–92.

Dammak, I., R. A. de Carvalho, C. S. F. Trindade, R. V. Lourenço and P. J. Do Amaral Sobral (2017). "Properties of active gelatin films incorporated with rutin-loaded nanoemulsions". International journal of biological macromolecules **98**: 39–49.

Darbasi, M., G. Askari, H. Kiani and F. Khodaiyan (2017). "Development of chitosan based extended-release antioxidant films by control of fabrication variables". International Journal of Biological Macromolecules **104**: 303–310.

Dash, K. K., P. Deka, S. P. Bangar, V. Chaudhary, M. Trif and A. Rusu (2022). "Applications of Inorganic Nanoparticles in Food Packaging: A Comprehensive Review". Polymers **14**(3): 521, 1–14. doi.org/10.3390/polym14030521

de Carvalho, S. M., C. M. Noronha, C. G. da Rosa, W. G. Sganzerla, I. C. Bellettini, M. R. Nunes, F. C. Bertoldi and P. L. Manique Barreto (2019). "PVA antioxidant nanocomposite films functionalized with alpha-tocopherol loaded solid lipid nanoparticles". Colloids and surfaces. A, Physicochemical and engineering aspects **581**: 123793.

Dehghani, S., S. V. Hosseini and J. M. Regenstein (2018). "Edible films and coatings in seafood preservation: A review". Food chemistry **240**: 505–513.

Donsi, F. and G. Ferrari (2016). "Essential oil nanoemulsions as antimicrobial agents in food". Journal of Biotechnology **233**: 106–120.

Douglass, E. F., H. Avci, R. Boy, O. J. Rojas and R. Kotek (2018). "A Review of Cellulose and Cellulose Blends for Preparation of Bio-derived and Conventional Membranes, Nanostructured Thin Films, and Composites". Polymer Reviews **58**(1): 102–163.

Duan, C., X. Meng, J. Meng, M. I. H. Khan, L. Dai, A. Khan, X. An, J. Zhang, T. Huq and Y. Ni (2019). "Chitosan as A Preservative for Fruits and Vegetables: A Review on Chemistry and Antimicrobial Properties". Journal of Bioresources and Bioproducts **4**(1): 11–21.

Ellis, D. I., V. L. Brewster, W. B. Dunn, J. W. Allwood, A. P. Golovanov and R. Goodacre (2012). "Fingerprinting food: current technologies for the detection of food adulteration and contamination". Chemical Society Reviews **41**(17): 5706–5727.

Erciyes, A. and B. Ocak (2019). "Physico-mechanical, thermal, and ultraviolet light barrier properties of collagen hydrolysate films from leather solid wastes incorporated with nano TiO2". Polymer Composites **40**(12): 4716–4725.

Esfanjani, A. F., E. Assadpour and S. M. Jafari (2018). "Improving the bioavailability of phenolic compounds by loading them within lipid-based nanocarriers". Trends in Food Science & Technology **76**: 56–66.

Fathi, M., A. Martin and D. J. McClements (2014). "Nanoencapsulation of food ingredients using carbohydrate based delivery systems". Trends in Food Science & Technology **39**(1): 18–39.

Fathi, M., M. R. Mozafari and M. Mohebbi (2012). "Nanoencapsulation of food ingredients using lipid based delivery systems". Trends in Food Science & Technology **23**(1): 13–27.

Fathi, N., H. Almasi and M. K. Pirouzifard (2018). "Effect of ultraviolet radiation on morphological and physicochemical properties of sesame protein isolate based edible films". Food Hydrocolloids **85**: 136–143.

Ferrentino, G., K. Morozova, C. Horn and M. Scampicchio (2020). "Extraction of Essential Oils from Medicinal Plants and their Utilization as Food Antioxidants". Current Pharmaceutical Design **26**(5): 519–541.

Ghorbani, F. M., B. Kaffashi, P. Shokrollahi, E. Seyedjafari and A. Ardeshirylajimi (2015). "PCL/chitosan/Zn-doped nHA electrospun nanocomposite scaffold promotes adipose derived stem cells adhesion and proliferation". Carbohydrate Polymers **118**: 133–142.

Gomes, A. and P. J. D. Sobral (2022). "Plant Protein-Based Delivery Systems: An Emerging Approach for Increasing the Efficacy of Lipophilic Bioactive Compounds". Molecules **27**(1): 60, 1–15.

Grillo, R., B. D. Mattos, D. R. Antunes, M. M. L. Forini, F. A. Monikh and O. J. Rojas (2021). "Foliage adhesion and interactions with particulate delivery systems for plant nanobionics and intelligent agriculture". Nano Today **37**: 101078, 1–9. doi.org/10.1016/j.nantod.2021.101078

Gutierrez-del-Rio, I., J. Fernandez and F. Lombo (2018). "Plant nutraceuticals as antimicrobial agents in food preservation: terpenoids, polyphenols and thiols". International Journal of Antimicrobial Agents **52**(3): 309–315.

Gutierrez-del-Rio, I., S. Lopez-Ibanez, P. Magadan-Corpas, L. Fernandez-Calleja, A. Perez-Valero, M. Tunon-Granda, E. M. Miguelez, C. J. Villar and F. Lombo (2021). "Terpenoids and Polyphenols as Natural Antioxidant Agents in Food Preservation". Antioxidants **10**(8): 1264, 1–12. https://doi.org/10.3390/antiox10081264

Hassoun, A., M. Carpena, M. A. Prieto, J. Simal-Gandara, F. Özogul, Y. Özogul, Ö. E. Çoban, M. Guðjónsdóttir, F. J. Barba, F. J. Marti-Quijal, A. R. Jambrak, N. Maltar-Strmečki, J. G. Kljusurić and J. M. Regenstein (2020). "Use of Spectroscopic Techniques to Monitor Changes in Food Quality during Application of Natural Preservatives: A Review". Antioxidants **9**(9): 882, 1–15. doi.org/10.3390/antiox9090882.

Helanto, K., L. Matikainen, R. Talja and O. J. Rojas (2019). "Bio-based Polymers for Sustainable Packaging and Biobarriers: A Critical Review". Bioresources **14**(2): 4902–4951.

Homayonpour, P., H. Jalali, N. Shariatifar and M. Amanlou (2021). "Effects of nano-chitosan coatings incorporating with free /nano encapsulated cumin (Cuminum cyminum L.) essential oil on quality characteristics of sardine fillet". International Journal of Food Microbiology **341**: 109047.

Hosseini, S. F., M. Rezaei, M. Zandi and F. Farahmandghavi (2016). "Development of bioactive fish gelatin/chitosan nanoparticles composite films with antimicrobial properties". Food Chemistry **194**: 1266–1274.

Huang, Y. Q., X. H. Wang, K. Q. Lai, Y. X. Fan and B. A. Rasco (2020). "Trace analysis of organic compounds in foods with surface-enhanced Raman spectroscopy: Methodology, progress, and challenges". Comprehensive Reviews in Food Science and Food Safety **19**(2): 622–642.

Jafarzadeh, S. and S. M. Jafari (2021). "Impact of metal nanoparticles on the mechanical, barrier, optical and thermal properties of biodegradable food packaging materials". Critical Reviews in Food Science and Nutrition **61**(16): 2640–2658.

Jafarzadeh, S., S. M. Jafari, A. Salehabadi, A. M. Nafchi, U. S. Uthaya Kumar and H. P. S. A. Khalil (2020). "Biodegradable green packaging with antimicrobial functions based on the bioactive compounds from tropical plants and their by-products". Trends in Food Science & Technology **100**: 262–277.

Jafarzadeh, S., J. W. Rhim, A. Alias, F. Ariffin and S. Mahmud (2019). "Application of antimicrobial active packaging film made of semolina flour, nano zinc oxide and nano-kaolin to maintain the quality of low-moisture mozzarella cheese during low-temperature storage". Journal of the Science of Food and Agriculture **99**(6): 2716–2725.

Jiang, L., M. M. Hassan, S. Ali, H. H. Li, R. Sheng and Q. S. Chen (2021). "Evolving trends in SERS-based techniques for food quality and safety: A review". Trends in Food Science & Technology **112**: 225–240.

Kalia, A., M. Kaur, A. Shami, S. K. Jawandha, M. A. Alghuthaymi, A. Thakur and K. A. Abd-Elsalam (2021). "Nettle-Leaf Extract Derived ZnO/CuO Nanoparticle-Biopolymer-Based Antioxidant and Antimicrobial Nanocomposite Packaging Films and Their Impact on Extending the Post-Harvest Shelf Life of Guava Fruit". Biomolecules **11**(2): 224, 1–12. doi.org/10.3390/biom11020224

Kamkar, A., E. Molaee-Aghaee, A. Khanjari, A. Akhondzadeh-Basti, B. Noudoost, N. Shariatifar, M. A. Sani and M. Soleimani (2021). "Nanocomposite active packaging based on chitosan biopolymer loaded with nano-liposomal essential oil: Its characterizations and effects on microbial, and chemical properties of refrigerated chicken breast fillet." International Journal of Food Microbiology **342**: 109071, 1–12. doi.org/10.1016/j.ijfoodmicro.2021.109071

Kaschuk, J. J., Y. Al Haj, O. J. Rojas, K. Miettunen, T. Abitbol and J. Vapaavuori (2022). "Plant-Based Structures as an Opportunity to Engineer Optical Functions in Next-Generation Light Management". Advanced Materials **34**(6): 2104473, 1–16. doi.org/10.1002/adma.202104473

Kazemeini, H., A. Azizian and H. Adib (2021). "Inhibition of Listeria monocytogenes growth in turkey fillets by alginate edible coating with Trachyspermum ammi essential oil nano-emulsion." International Journal of Food Microbiology **344**: 109104, 1–18. doi.org/https://doi.org/10.1016/j.ijfoodmicro.2021.109104

Khezerlou, A., A. Ehsani, M. Tabibiazar and E. Moghaddas Kia (2019). "Development and characterization of a Persian gum–sodium caseinate biocomposite film accompanied by Zingiber officinale extract". Journal of Applied Polymer Science **136**(12): 47215, 19. doi.org/10.1002/app.47215

Khezerlou, A., H. Zolfaghari, S. A. Banihashemi, S. Forghani and A. Ehsani (2021). "Plant gums as the functional compounds for edible films and coatings in the food industry: A review". Polymers for Advanced Technologies **32**(6): 2306–2326.

Kocabas, D. S., M. E. Akcelik, E. Bahcegul and H. N. Ozbek (2021). "Bulgur bran as a biopolymer source: Production and characterization of nanocellulose-reinforced hemicellulose-based biodegradable films with decreased water solubility". Industrial Crops and Products **171**: 113847, 1–14. doi.org/10.1016/j.indcrop.2021.113847

Kotsanopoulos, K. V. and I. S. Arvanitoyannis (2015). "Membrane Processing Technology in the Food Industry: Food Processing, Wastewater Treatment, and Effects on Physical, Microbiological, Organoleptic, and Nutritional Properties of Foods". Critical Reviews in Food Science and Nutrition **55**(9): 1147–1175.

Krishna, V. D., K. Wu, D. Q. Su, M. C. J. Cheeran, J. P. Wang and A. Perez (2018). "Nanotechnology: Review of concepts and potential application of sensing platforms in food safety". Food Microbiology **75**: 47–54.

Kumar, T. S. M., K. S. Kumar, N. Rajini, S. Siengchin, N. Ayrilmis and A. V. Rajulu (2019). "A comprehensive review of electrospun nanofibers: Food and packaging perspective". Composites Part B-Engineering **175**: 107074, 1–20. doi.org/10.1016/j.compositesb.2019.107074

Lambert, A. D., J. P. Smith and K. L. Dodds (1991). "Shelf life extension and microbiological safety of fresh meat – a review". Food Microbiology **8**(4): 267–297.

Lei, K., X. Wang, X. Li and L. Wang (2019). "The innovative fabrication and applications of carvacrol nanoemulsions, carboxymethyl chitosan microgels and their composite films". Colloids and surfaces. B, Biointerfaces **175**: 688–696.

Li, C., X.-L. Qiu, L.-X. Lu, Y.-L. Tang, Q. Long and J.-G. Dang (2019a). "Preparation of low-density polyethylene film with quercetin and α-tocopherol loaded with mesoporous silica for synergetic-release antioxidant active packaging". Journal of Food Process Engineering **42**(5): e13088.

Li, X., X. Yang, H. Deng, Y. Guo and J. Xue (2020). "Gelatin films incorporated with thymol nanoemulsions: Physical properties and antimicrobial activities". International Journal of Biological Macromolecules **150**: 161–168.

Li, Y., Z. X. Wang, L. Sun, L. Q. Liu, C. L. Xu and H. Kuang (2019b). "Nanoparticle-based sensors for food contaminants". Trac-Trends in Analytical Chemistry **113**: 74–83.

Lima, R. C., A. P. A. de Carvalho, C. P. Vieira, R. V. Moreira and C. A. Conte (2021). "Green and Healthier Alternatives to Chemical Additives as Cheese Preservative: Natural Antimicrobials in Active Nanopackaging/Coatings". Polymers **13**(16): 2675, 1–12. doi.org/10.3390/polym13162675

Liu, G. Y., M. Lu, X. D. Huang, T. F. Li and D. H. Xu (2018). "Application of Gold-Nanoparticle Colorimetric Sensing to Rapid Food Safety Screening". Sensors **18**(12): 4166, 1–12. doi.org/10.3390/s18124166

Liu, Y. X., H. H. Liu and Z. R. Shen (2021). "Nanocellulose Based Filtration Membrane in Industrial Waste Water Treatment: A Review". Materials **14**(18): 5398, 1–18. doi.org/10.3390/ma14185398

Maqsoudlou, A., E. Assadpour, H. Mohebodini and S. M. Jafari (2020). "Improving the efficiency of natural antioxidant compounds via different nanocarriers." Advances in Colloid and Interface Science **278**: 102122, 1–19. doi.org/https://doi.org/10.1016/j.cis.2020.102122.

Matthews, K. R., K. E. Kniel and T. J. Montville (2017). Food Microbiology: An Introduction. Washington, D.C., ASM Press.

Mautner, A. (2020). "Nanocellulose water treatment membranes and filters: a review". Polymer International **69**(9): 741–751.

McClements, D. J. (2015). Nanoparticle- and Microparticle-based Delivery Systems. Boca Raton, FL, CRC Press.

McClements, D. J. and B. Ozturk (2021). "Utilization of Nanotechnology to Improve the Handling, Storage and Biocompatibility of Bioactive Lipids in Food Applications". Foods **10**(2): 365, 1–15. doi.org/10.3390/foods10020365

McClements, D. J. and J. Rao (2011). "Food-Grade Nanoemulsions: Formulation, Fabrication, Properties, Performance, Biological Fate, and Potential Toxicity". Critical Reviews in Food Science and Nutrition **51**(4): 285–330.

Mirzaei-Mohkam, A., F. Garavand, D. Dehnad, J. Keramat and A. Nasirpour (2020). "Physical, mechanical, thermal and structural characteristics of nanoencapsulated vitamin E loaded carboxymethyl cellulose films." Progress in Organic Coatings **138**: 105383, 1–9. doi.org/https://doi.org/10.1016/j.porgcoat.2019.105383.

Motelica, L., D. Ficai, O. C. Oprea, A. Ficai, V. L. Ene, B. S. Vasile, E. Andronescu and A. M. Holban (2021). "Antibacterial Biodegradable Films Based on Alginate with Silver Nanoparticles and Lemongrass Essential Oil–Innovative Packaging for Cheese". Nanomaterials **11**(9): 2377, 1–11. doi.org/10.3390/nano11092377

Mustafa, F. and S. Andreescu (2018). "Chemical and Biological Sensors for Food-Quality Monitoring and Smart Packaging". Foods **7**(10): 168, 1–15. https://doi.org/10.3390/foods7100168.

Naghizadeh, A. and M. Ghofouri (2019). "Synthesis of Low Cost Nanochitosan from Persian Gulf Shrimp Shell for Efficient Removal of Reactive Blue 29 (RB29) Dye from Aqueous Solution". Iranian Journal of Chemistry & Chemical Engineering-International English Edition **38**(6): 93–103.

Nair, S. B., N. J. Alummoottil and S. S. Moothandasserry (2017). "Chitosan-konjac glucomannan-cassava starch-nanosilver composite films with moisture resistant and antimicrobial properties for food-packaging applications". Starch-Starke **69**: 1–2.

Nazari, M., H. Majdi, M. Milani, S. Abbaspour-Ravasjani, H. Hamishehkar and L.-T. Lim (2019). "Cinnamon nanophytosomes embedded electrospun nanofiber: Its effects on microbial quality and shelf-life of shrimp as a novel packaging." Food Packaging and Shelf Life 21: 100349, 1–13. doi.org/10.1016/j.fpsl.2019.100349

Nejad, F. G., S. Tajik, H. Beitollahi and I. Sheikhshoaie (2021). "Magnetic nanomaterials based electrochemical (bio)sensors for food analysis". Talanta 228: 122075, 1–15. doi.org/10.1016/j.talanta.2020.122075

Nejatian, M., H. Saberian and S. M. Jafari (2019). Encapsulation of food ingredients by double nanoemulsions. Lipid-Based Nanostructures for Food Encapsulation Purposes, Elsevier: 89–128.

Nielsen, S. S. (2017). Food Analsysis. New York, NY, Springer.

Niu, B., Z. P. Yan, P. Shao, J. Kang and H. J. Chen (2018). "Encapsulation of Cinnamon Essential Oil for Active Food Packaging Film with Synergistic Antimicrobial Activity". Nanomaterials 8(8): 598, 1–15. https://doi.org/10.3390/nano8080598.

Noshirvani, N., B. Ghanbarzadeh, R. R. Mokarram and M. Hashemi (2017). "Novel active packaging based on carboxymethyl cellulose-chitosan -ZnO NPs nanocomposite for increasing the shelf life of bread". Food Packaging and Shelf Life 11: 106–114.

Otoni, C. G., H. M. C. Azeredo, B. D. Mattos, M. Beaumont, D. S. Correa and O. J. Rojas (2021). "The Food-Materials Nexus: Next Generation Bioplastics and Advanced Materials from Agri-Food Residues". Advanced Materials 33(43): 2102520, 1–24. https://doi.org/10.1002/adma.202102520.

Otoni, C. G., S. F. Pontes, E. A. Medeiros and N. D. F. Soares (2014). "Edible films from methylcellulose and nanoemulsions of clove bud (Syzygium aromaticum) and oregano (Origanum vulgare) essential oils as shelf life extenders for sliced bread". Journal of Agricultural and Food Chemistry 62(22): 5214–5219.

Papadimitriou, A., I. Ketikidis, M. E. K. Stathopoulou, C. N. Banti, C. Papachristodoulou, L. Zoumpoulakis, S. Agathopoulos, G. V. Vagenas and S. K. Hadjikakou (2018). "Innovative material containing the natural product curcumin, with enhanced antimicrobial properties for active packaging". Materials Science and Engineering: C 84: 118–122.

Pérez-Córdoba, L. J., I. T. Norton, H. K. Batchelor, K. Gkatzionis, F. Spyropoulos and P. J. Sobral (2018). "Physico-chemical, antimicrobial and antioxidant properties of gelatin-chitosan based films loaded with nanoemulsions encapsulating active compounds". Food Hydrocolloids 79: 544–559.

Petersen, M., Z. L. Yu and X. N. Lu (2021). "Application of Raman Spectroscopic Methods in Food Safety: A Review". Biosensors-Basel 11(6): 187, 1–18. https://doi.org/10.3390/bios11060187.

Pirouzifard, M., R. A. Yorghanlu and S. Pirsa (2020). "Production of active film based on potato starch containing Zedo gum and essential oil of Salvia officinalis and study of physical, mechanical, and antioxidant properties". Journal of Thermoplastic Composite Materials 33(7): 915–937.

Pristouri, G., A. Badeka and M. G. Kontominas (2010). "Effect of packaging material headspace, oxygen and light transmission, temperature and storage time on quality characteristics of extra virgin olive oil". Food Control 21(4): 412–418.

Qazanfarzadeh, Z. and M. Kadivar (2016). "Properties of whey protein isolate nanocomposite films reinforced with nanocellulose isolated from oat husk". International Journal of Biological Macromolecules 91: 1134–1140.

Ranjbaryan, S., B. Pourfathi and H. Almasi (2019). "Reinforcing and release controlling effect of cellulose nanofiber in sodium caseinate films activated by nanoemulsified cinnamon essential oil". Food Packaging and Shelf Life 21: 100341.

Rao, J. J., B. C. Chen and D. J. McClements (2019). "Improving the Efficacy of Essential Oils as Antimicrobials in Foods: Mechanisms of Action". Annual Review of Food Science and Technology Vol 10. M. P. Doyle and D. J. McClements 10: 365–387.

Rezaei, A., M. Fathi and S. M. Jafari (2019). "Nanoencapsulation of hydrophobic and low-soluble food bioactive compounds within different nanocarriers". Food Hydrocolloids **88**: 146–162.

Sahraee, S., J. M. Milani, J. M. Regenstein and H. S. Kafil (2019). "Protection of foods against oxidative deterioration using edible films and coatings: A review". Food Bioscience **32**: 100451.

Sami, R., E. Khojah, A. Elhakem, N. Benajiba, M. Helal, N. Alhuthal, S. A. Alzahrani, M. Alharbi and M. Chavali (2021). "Performance Study of Nano/SiO2 Films and the Antimicrobial Application on Cantaloupe Fruit Shelf-Life". Applied Sciences-Basel **11**(9): 3879, 1–9. doi.org/10.3390/app11093879

Sanchez-Gonzalez, L., M. Cháfer, M. Hernández, A. Chiralt and C. González-Martínez (2011). "Antimicrobial activity of polysaccharide films containing essential oils". Food Control **22**(8): 1302–1310.

Sani, M. A., M. Azizi-Lalabadi, M. Tavassoli, K. Mohammadi and D. J. McClements (2021a). "Recent Advances in the Development of Smart and Active Biodegradable Packaging Materials". Nanomaterials **11**(5): 1331, 1–9. https://doi.org/10.3390/nano11051331.

Sani, M. A., M. Tavassoli, H. Hamishehkar and D. J. McClements (2021b). "Carbohydrate-based films containing pH-sensitive red barberry anthocyanins: Application as biodegradable smart food packaging materials". Carbohydrate Polymers **255**: 117488, 1–12. doi.org/10.1016/j.carbpol.2020.117488

Sarkar, P., R. Choudhary, S. Panigrahi, I. Syed, S. Sivapratha and C. V. Dhumal (2017). "Nano-inspired systems in food technology and packaging". Environmental Chemistry Letters **15**(4): 607–622.

Scotter, M. (2009). "The chemistry and analysis of annatto food colouring: a review". Food Additives & Contaminants: Part A **26**(8): 1123–1145.

Seow, Y. X., C. R. Yeo, H. L. Chung and H. G. Yuk (2014). "Plant Essential Oils as Active Antimicrobial Agents". Critical Reviews in Food Science and Nutrition **54**(5): 625–644.

Sharifimehr, S., N. Soltanizadeh and S. A. Hossein Goli (2019). "Effects of edible coating containing nano-emulsion of Aloe vera and eugenol on the physicochemical properties of shrimp during cold storage". Journal of the Science of Food and Agriculture **99**(7): 3604–3615.

Sharma, P. R., S. K. Sharma, T. Lindstrom and B. S. Hsiao (2020). "Nanocellulose-Enabled Membranes for Water Purification: Perspectives". Advanced Sustainable Systems **4**(5): 1900114, 1–14. doi.org/10.1002/adsu.201900114

Shen, Y., Z.-J. Ni, K. Thakur, J.-G. Zhang, F. Hu and Z.-J. Wei (2021). "Preparation and characterization of clove essential oil loaded nanoemulsion and pickering emulsion activated pullulan-gelatin based edible film". International Journal of Biological Macromolecules **181**: 528–539.

Shewan, H. M. and J. R. Stokes (2013). "Review of techniques to manufacture micro-hydrogel particles for the food industry and their applications". Journal of Food Engineering **119**(4): 781–792.

Siddiqui, J., M. Taheri, A. Ul Alam and M. J. Deen (2022). "Nanomaterials in Smart Packaging Applications: A Review". Small **18**(1): 2101171, 1–13. https://doi.org/10.1002/smll.202101171.

Silva, H. D., M. A. Cerqueira and A. A. Vicente (2012). "Nanoemulsions for Food Applications: Development and Characterization". Food and Bioprocess Technology **5**(3): 854–867.

Stoll, L., S. Domenek, S. Hickmann Flôres, S. M. B. Nachtigall and A. de Oliveira Rios (2021). "Polylactide films produced with bixin and acetyl tributyl citrate: Functional properties for active packaging". Journal of Applied Polymer Science **138**(17): 50302, 1–9. doi.org/https://doi.org/10.1002/app.50302.

Sun, J., J. Shen, S. Chen, M. A. Cooper, H. Fu, D. Wu and Z. Yang (2018). "Nanofiller Reinforced Biodegradable PLA/PHA Composites: Current Status and Future Trends". Polymers **10**(5): 505, 1–11. doi.org/10.3390/polym10050505

Sun, X. D. and R. A. Holley (2012). "Antimicrobial and Antioxidative Strategies to Reduce Pathogens and Extend the Shelf Life of Fresh Red Meats". Comprehensive Reviews in Food Science and Food Safety **11**(4): 340–354.

Taherkhani, E., M. Moradi, H. Tajik, R. Molaei and P. Ezati (2020). "Preparation of on-package halochromic freshness/spoilage nanocellulose label for the visual shelf life estimation of meat". International Journal of Biological Macromolecules **164**: 2632–2640.

Tamjidi, F., M. Shahedi, J. Varshosaz and A. Nasirpour (2013). "Nanostructured lipid carriers (NLC): A potential delivery system for bioactive food molecules". Innovative Food Science & Emerging Technologies **19**: 29–43.

Tavassoli, M., M. A. Sani, A. Khezerlou, A. Ehsani and D. J. McClements (2021). "Multifunctional nanocomposite active packaging materials: Immobilization of quercetin, lactoferrin, and chitosan nanofiber particles in gelatin films". Food Hydrocolloids **118**: 106747.

Teixeira, R. F., C. A. Balbinot and C. D. Borges (2022). "Essential oils as natural antimicrobials for application in edible coatings for minimally processed apple and melon: A review on antimicrobial activity and characteristics of food models". Food Packaging and Shelf Life **31**: 100781, 1–15. https://doi.org/10.1016/j.fpsl.2021.100781.

Tonyali, B., A. McDaniel, J. Amamcharla, V. Trinetta and U. Yucel (2020). "Release kinetics of cinnamaldehyde, eugenol, and thymol from sustainable and biodegradable active packaging films." Food Packaging and Shelf Life **24**: 100484, 1–9. doi.org/10.1016/j.fpsl.2020.100484.

Tsironi, T. N. and P. S. Taoukis (2018). "Current Practice and Innovations in Fish Packaging". Journal of Aquatic Food Product Technology **27**(10): 1024–1047.

Valdés, A., A. C. Mellinas, M. Ramos, N. Burgos, A. Jiménez and M. C. Garrigós (2015). "Use of herbs, spices and their bioactive compounds in active food packaging". RSC Advances **5**(50): 40324–40335.

Wang, K. Q., Z. L. Li, J. J. Li and H. Lin (2021). "Raman spectroscopic techniques for nondestructive analysis of agri-foods: A state-of-the-art review". Trends in Food Science & Technology **118**: 490–504.

Wang, Y. and T. V. Duncan (2017). "Nanoscale sensors for assuring the safety of food products". Current Opinion in Biotechnology **44**: 74–86.

Wei, N. S., M. R. Liao, K. J. Xu and Z. Y. Qin (2021). "High-performance soy protein-based films from cellulose nanofibers and graphene oxide constructed synergistically via hydrogen and chemical bonding". Rsc Advances **11**(37): 22812–22819.

Weiss, J., E. A. Decker, D. J. McClements, K. Kristbergsson, T. Helgason and T. Awad (2008). "Solid lipid nanoparticles as delivery systems for bioactive food components". Food Biophysics **3**(2): 146–154.

Wink, M. (2022). "Current Understanding of Modes of Action of Multicomponent Bioactive Phytochemicals: Potential for Nutraceuticals and Antimicrobials". Annual Review of Food Science and Technology **13**(1): 337–359.

Xavier, L. O., W. G. Sganzerla, G. B. Rosa, C. G. da Rosa, L. Agostinetto, A. P. D. L. Veeck, L. C. Bretanha, G. A. Micke, M. Dalla Costa, F. C. Bertoldi, P. L. M. Barreto and M. R. Nunes (2021). "Chitosan packaging functionalized with Cinnamodendron dinisii essential oil loaded zein: A proposal for meat conservation". International Journal of Biological Macromolecules **169**: 183–193.

Xiao, Y. Q., Y. N. Liu, S. F. Kang, K. H. Wang and H. D. Xu (2020). "Development and evaluation of soy protein isolate-based antibacterial nanocomposite films containing cellulose nanocrystals and zinc oxide nanoparticles". Food Hydrocolloids **106**: 105898, 1–10. doi.org/10.1016/j.foodhyd.2020.105898

Xiong, Y., S. Li, R. D. Warner and Z. Fang (2020). "Effect of oregano essential oil and resveratrol nanoemulsion loaded pectin edible coating on the preservation of pork loin in modified

atmosphere packaging." Food Control **114**: 107226, 1–12. doi.org/10.1016/j.
foodcont.2020.107226

Yadav, S., G. K. Mehrotra, P. Bhartiya, A. Singh and P. K. Dutta (2020). "Preparation,
physicochemical and biological evaluation of quercetin based chitosan-gelatin film for food
packaging." Carbohydrate Polymers **227**: 115348, 1–11. doi.org/10.1016/j.carbpol.2019.115348

Yadav, S., G. K. Mehrotra and P. K. Dutta (2021). "Chitosan based ZnO nanoparticles loaded gallic-
acid films for active food packaging". Food Chemistry **334**: 127605, 1–13. doi.org/10.1016/j.
foodchem.2020.127605

Yang, Y. Q., N. Creedon, A. O'Riordan and P. Lovera (2021). "Surface Enhanced Raman
Spectroscopy: Applications in Agriculture and Food Safety". Photonics **8**(12): 568, 1–16.

Zambrano-Zaragoza, M. L., D. Quintanar-Guerrero, A. Del Real, E. Piñon-Segundo and
J. F. Zambrano-Zaragoza (2017). "The release kinetics of β-carotene nanocapsules/xanthan
gum coating and quality changes in fresh-cut melon (cantaloupe)". Carbohydrate Polymers
157: 1874–1882.

Zhai, X. D., X. B. Zou, J. Y. Shi, X. W. Huang, Z. B. Sun, Z. H. Li, Y. Sun, Y. X. Li, X. Wang, M. Holmes,
Y. Y. Gong, M. Povey and J. B. Xiao (2020). "Amine-responsive bilayer films with improved
illumination stability and electrochemical writing property for visual monitoring of meat
spoilage". Sensors and Actuators B-Chemical **302**: 127130, 1–12. doi.org/10.1016/j.
snb.2019.127130

Zheng, J. K. and L. L. He (2014). "Surface-Enhanced Raman Spectroscopy for the Chemical Analysis
of Food". Comprehensive Reviews in Food Science and Food Safety **13**(3): 317–328.

Chapter 6
Nanotoxicology: The potential risks of food nanotechnology

6.1 Introduction

One of the main concerns with using nanotechnology in foods is the potential risk to human health of consuming nanomaterials as well as damage to the environment (More et al. 2021). As discussed earlier, one of the main reasons nanoparticles are utilized in the food industry is due to their unusual properties, which are associated with their relatively small dimensions. For instance, small particles are digested more rapidly, have a higher surface reactivity, and can penetrate through biological barriers more effectively than larger ones. These characteristics are useful to obtain novel or improved properties in foods, but they could also have unforeseen adverse effects on human health (Sharifi et al. 2012, McClements and Xiao 2017). Consequently, it is critical for those working in this area to consider both the risks and benefits of introducing nanoparticles into foods (More et al. 2021). Concerns about the potential adverse effects of nanomaterials on human health and the environment may arise in several areas:

–   *Production*: There may be concerns about the toxicity of nanoparticles in manufacturing facilities, especially when they are produced in a powdered form. If nanoparticles get into the air, then they may be taken into the lungs of personnel working in the facility by inhalation.
–   *Intentional ingestion*: Some food components are natural nanoparticles or are intentionally designed to have nanoscale dimensions, such as casein micelles, small oil bodies, nanoemulsion droplets, and titanium dioxide nanoparticles. These nanoparticles will enter the human body *via* the gastrointestinal tract (Figure 6.1).
–   *Unintentional ingestion*: Nanoparticles may enter foods without being intentionally added. For instance, they may contaminate the water or ingredients used to formulate a food, or they may leach into foods from packaging materials. Again, these nanoparticles will enter the body by the gastrointestinal route (Figure 6.1).
–   *Environmental contamination*: The nanopesticides or nanofertilizers used to treat agricultural crops may contaminate foods, or they may leach into the environment and enter the soil and the food chain. Moreover, nanoparticles in packaging materials or within foods may migrate into the environment when these materials are disposed.

In this chapter, the focus will mainly be on nanoparticles that enter the gastrointestinal tract since these are the ones most likely to impact human health and wellbeing through foods. Initially, the different kinds of inorganic and organic nanoparticles

https://doi.org/10.1515/9783110788457-006

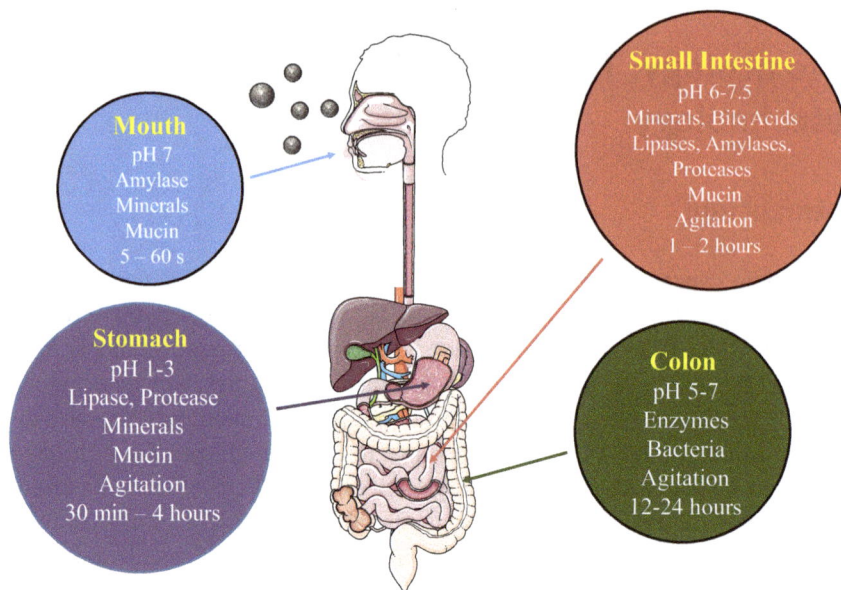

Figure 6.1: Schematic diagram of the physicochemical and physiological conditions in different regions of the human gastrointestinal tract. The diagram of the human body is from Servier Medical Art (Creative Commons 3.0).

that may be intentionally or unintentionally ingested are introduced and then the impact of nanoparticle properties on their potential behavior in the human gastrointestinal tract is discussed as well as the importance of food matrix effects. The potential physicochemical or physiological mechanisms that may lead to the toxicity of nanoparticles in foods are discussed.

6.2 Kinds of nanoparticles present in foods

As discussed in Chapter 1, both inorganic and organic nanoparticles may be found in foods. The nature of these nanoparticles plays an important role in determining their behavior in the gastrointestinal tract as well as their potential to cause toxicity. For this reason, we begin by discussing the characteristics of these two main categories of food nanoparticles.

6.2.1 Inorganic nanoparticles

Inorganic nanoparticles, such as those composed of silver, zinc oxide (ZnO), iron oxide, titanium dioxide, or silicon dioxide (SiO_2), are currently being utilized or are being explored for their potential application within foods and food packaging materials (Pietroiusti et al. 2016). At room temperature, inorganic nanoparticles may have different physical states (crystalline or amorphous), shapes (e.g., spherical, cuboid, fibrous, or irregular), sizes (5–1000 nm), and interfacial properties (e.g., thickness, charge, hydrophobicity, and reactivity), which depend on the materials and processing operations used to fabricate them. The propensity of inorganic nanoparticles to dissolve, disperse, or aggregate when exposed to specific solution conditions (e.g., pH, ionic strength, solvent type, and temperature), as well as their ability to participate in specific chemical reactions, influences their behavior in the gastrointestinal tract as well as their potential toxicity. In the remainder of this section, several of the most important types of inorganic nanoparticles used in foods are considered.

6.2.1.1 Silver nanoparticles

Silver nanoparticles have been utilized for several purposes in the food industry, especially because of their strong antimicrobial properties (Hajipour et al. 2012, Gaillet and Rouanet 2015, Pulit-Prociak et al. 2015). For instance, they have been incorporated into food packaging materials and containers to protect foods from contamination by spoilage or pathogenic microorganisms during storage (Echegoyen and Nerin 2013b). A concern with this application of inorganic nanoparticles is that they could migrate into the food matrix, which means they could be consumed by humans (Echegoyen and Nerin 2013b, Echegoyen and Nerin 2013a, Mackevica et al. 2016). Moreover, studies have shown that silver nanoparticles may form in food matrices or gastrointestinal fluids because soluble silver ions interact with other substances and precipitate (Loza et al. 2014). It has been reported that adults can ingest from around 20 to 80 µg/day of silver but only a small fraction of this would be as nanoparticles (Frohlich and Frohlich 2016). There is currently a relatively poor understanding of the toxicity of silver nanoparticles that may be consumed as part of foods or beverages (Gaillet and Rouanet 2015). Some studies suggest they are toxic whereas others suggest they are not, which is probably because of differences in the dose and type of silver nanoparticles, foods, and test methods employed in these studies. After oral ingestion by animals, silver nanoparticles have been reported to penetrate lymphocytes, causing a change in villi color, promoting mucus granule discharge, and altering mucus composition in the small intestine (Cha et al. 2008, Jeong et al. 2010, Kim et al. 2010b, Shahare and Yashpal 2013). Orally administered silver nanoparticles have been reported to accumulate within the liver, kidneys, spleen, stomach, and small intestine of animals (Kim et al. 2008, Gaillet and Rouanet 2015, Hendrickson et al. 2016). Taken together, these studies suggest that silver nanoparticles can be absorbed by the gastrointestinal tract and then distributed throughout different organs and tissues. Nevertheless, some studies have reported that only a low proportion (<1%) of orally administered silver nanoparticles actually accumulate in organs and tissues, suggesting that most of them are either not absorbed or are rapidly excreted

(Hendrickson et al. 2016). Moreover, the authors of this study reported no appreciable toxicity of the nanoparticles at the doses used, which were 2000 mg/kg body weight for a single dose or 250 mg/kg body weight for multiple doses (Hendrickson et al. 2016). No major toxicity was also reported in a different rat feeding study when different doses of silver nanoparticles (30, 300, or 1000 mg/kg) were orally administered to the animals over 28 days (Kim et al. 2008). However, this study did report that there was slight damage to the liver for the highest dose used. Repeated oral administration of silver nanoparticles to mice has also been reported to cause some adverse effects on liver and kidney function in the animals (Kim et al. 2010a, Park et al. 2010). Some researchers have proposed that adverse effects on liver function are observed when the level of silver nanoparticles consumed exceeds about 125 mg/kg body weight (Kim et al. 2010a).

To summarize, the results of animal studies suggest that silver nanoparticles can accumulate within the body and may exhibit some toxicity when ingested at sufficiently high concentrations. Nevertheless, it is still unclear whether the nanoparticle concentrations where adverse effects are observed are exceeded in real food applications. For this reason, it will be important to carry out long-term feeding studies on the potential toxicity of silver nanoparticles using concentrations that are relevant to those present in the human diet. Moreover, additional studies are needed to assess whether ingested silver nanoparticles fully dissolve within the fluids inside the gastrointestinal tract, whether ingested silver ions can form nanoparticles inside the gastrointestinal tract, and to establish the impact of soluble and insoluble forms of silver on their potential toxicity (Loeschner et al. 2011). In one study, it was reported that the distribution of silver within the organs of the animals was not strongly dependent on whether it was ingested in a soluble or nanoparticle form.

Cell culture models have been used to identify the potential mechanisms of cytotoxicity of silver nanoparticles and to establish the major factors that impact their adverse effects (Chen et al. 2016, Georgantzopoulou et al. 2016). These studies indicate that silver nanoparticle toxicity increases as their particle size decreases (Kim et al. 2012) and depends on their interfacial properties (Sharma et al. 2014). Silver nanoparticles have been reported to produce reactive oxygen species (ROS) within cellular environments, which puts the cells under oxidative stress and can cause damage to genetic materials, enzymes, membranes, and organelles (Sharma et al. 2014, Gaillet and Rouanet 2015). Silver nanoparticles can also interfere with important biochemical processes within cells, including the expression of genes, DNA replication, and the production of ATP (Sharma et al. 2014). They may also influence the composition of the microflora within the colon because of their antimicrobial activity, which could have negative effects on human health and wellbeing (Williams et al. 2015, Frohlich and Frohlich 2016). Some cell culture studies have shown that food matrix effects can influence the absorption and cytotoxicity of silver nanoparticles (Lichtenstein et al. 2015). It is important to note, however, that the nanoparticle doses used in cell culture studies are often much greater than that what would normally be found in the human diet. Moreover, food matrix effects are often ignored in these studies. Consequently, the results obtained may not be relevant to human ingestion of silver nanoparticles.

6.2.1.2 Zinc oxide

Nanoparticles composed of ZnO can be utilized as a bioavailable source of zinc in foods (Wang et al. 2014). Zinc is an essential mineral element that is required for healthy body function. ZnO nanoparticles have also been used as antimicrobials in food packaging materials to inhibit the growth of spoilage or pathogenic microbes (Sirelkhatim et al. 2015). They have also been incorporated into packaging materials to block ultraviolet light waves from entering and damaging photosensitive components in foods (EFSA 2016). As mentioned earlier, nanoparticles used as additives within packaging materials can migrate into foods and then be orally ingested (Bumbudsanpharoke and Ko 2015). Nevertheless, a risk assessment of this effect has reported that this phenomenon is not a major concern for ZnO nanoparticles (EFSA 2016). The antimicrobial activity of ZnO nanoparticles may be a result of their ability to penetrate microbial cells and form ROS that disrupt critical cellular constituents (such as genetic materials, enzymes, receptors, or membranes), which then leads to cytotoxicity (Sirelkhatim et al. 2015). Studies where ZnO nanoparticles are orally administered to rodents have shown that their absorption usually increases as their size decreases (Wang et al. 2008, Pasupuleti et al. 2012, Vandebriel and De Jong 2012, Wang et al. 2013a, Liang et al. 2017). Cell culture models have also shown that the absorption and toxicity of this kind of nanoparticle tend to increase as the particle size decreases (Mittag et al. 2021). The effects of ZnO nanoparticle dose have also been investigated. For instance, a study showed that oral administration of high levels of ZnO nanoparticles (50 or 100 mg/kg) to rats caused intestinal injury but lower levels (5 mg/kg) did not (Abbasi-Oshaghi et al. 2018). Some researchers have reported that a single oral dose of ZnO nanoparticles led to damage of the liver, kidney, and lungs of rodents (Esmaeillou et al. 2013). Another study reported that oral administration of ZnO nanoparticles alone did not lead to appreciable toxicity, but when they were administered with ascorbic acid, they did cause toxicity (Wang et al. 2014). This study highlights the importance of accounting for food matrix effects when establishing the potential toxicity of ingested nanoparticles. Indeed, researchers have shown that the potential toxicity of ZnO nanoparticles is influenced by the nature of the surrounding food matrix (Huang et al. 2019). For instance, delivering these nanoparticles in an acidic environment (e.g., with citric or ascorbic acid) increased their toxicity, which was related to the release of zinc ions (Zn^{2+}) from the nanoparticles in the presence of acids. Conversely, the toxicity of the ZnO nanoparticles was reduced when they were delivered with phosphate ions. In another study, it was shown that ZnO nanoparticles could interact with proteins in their environment but that these interactions did not impact their toxicity (which was relatively low), as determined by oral administration to rats for 14 days (Jung et al. 2021).

The size, shape, and aggregation state of ZnO nanoparticles vary considerably depending on how they are prepared and the prevailing environmental and solution conditions (Wang et al. 2014, Kang et al. 2015). Aggregated nanoparticles would be expected to behave differently within the gastrointestinal tract than non-aggregated ones, which would be expected to influence their potential toxicity. Oral administration of ZnO in soluble and nanoparticle forms to frogs showed that the nanoparticle form exhibited greater toxicity than the dissolved one (Bacchetta et al. 2014). It was suggested that the nanoparticle form was more effective at generating ROS and therefore damaging vital cellular components. The results of this research indicate the importance of determining the physical form of ingested zinc (soluble or particulate).

6.2.1.3 Iron oxide nanoparticles
Nanoparticles composed of iron oxide (Fe_2O_3) are used as pigments or nutrients in food products (Hilty et al. 2010, Raspopov et al. 2011, Zimmermann and Hilty 2011, Wu et al. 2014, Voss et al. 2020). In the USA, iron oxide is permitted for use up to 0.1% in some foods (such as sausage casings) as a colorant (WHO 2000). Studies suggest that on average people consume about 450 µg of iron oxide per day as part of their diet (WHO 2000), but these levels may be considerably higher for people who consume iron-fortified foods or supplements (Fulgoni et al. 2011). For instance, the ingestion of iron has been reported to range from around 10000 to 23000 µg per day from fortified foods and from around 10000 to 32000 µg per day for supplements (Fulgoni et al. 2011). However, most of this iron would not usually be consumed in the form of iron oxide nanoparticles.

Studies have shown that iron oxide nanoparticles tend to remain intact when they pass through simulated gastrointestinal conditions (Voss et al. 2021). The same authors showed that there was little uptake of these nanoparticles by model epithelium cells (Caco 2 cells) in a cell culture model. However, a study where these nanoparticles were orally administered to rodents showed that some of them were absorbed by the body without causing significant tissue damage (Garcia-Fernandez et al. 2020). A review of the potential interaction mechanisms of ZnO with the human body highlighted the fact that these nanoparticles are usually coated by a layer of proteins and other molecules after they enter the gastrointestinal tract leading to the formation of a biomolecular corona that alters their subsequent gastrointestinal fate and toxicity (Frtus et al. 2020). It was reported that the uptake of iron oxide nanoparticles by cells depends on their size, shape, aggregation state, and interfacial properties (Frtus et al. 2020). Consequently, their toxicity is likely to depend on the initial characteristics of the nanoparticles used as well as their interactions with food matrix and gastrointestinal constituents (Figure 6.2) (Patil et al.

2015). The ability of iron oxide nanoparticles to generate ROS has been proposed to be one of the primary factors contributing to their cytotoxicity (Wu et al. 2014, Frtus et al. 2020). As mentioned earlier, ROS can damage DNA, proteins, and lipids in cells, which could lead to oxidative stress and cell death. Even so, several studies have reported that oral administration of iron oxide nanoparticles did not promote accumulation or toxicity over a wide range of doses. For instance, oral administration of iron oxide nanoparticles to rats at levels ranging from 3 to 1000 mg/kg body weight for 13 weeks was not reported to lead to tissue accumulation or significant toxicity (Hilty et al. 2010, Yun et al. 2015). Conversely, other researchers reported that oral administration of iron oxide nanoparticles to rats caused liver damage at higher doses (150 and 300 mg/kg) but not at lower ones (25, 50, and 75 mg/kg) (Parivar et al. 2016). Clearly, there is still a need for further research to establish the potential impact of the ingestion of this type of nanoparticle on human health.

6.2.1.4 Titanium dioxide nanoparticles

Powdered titanium dioxide has been used as an additive in some foods for many years to improve their appearance by increasing their whiteness or brightness (Weir et al. 2012a). The TiO_2 ingredients developed for this purpose are usually designed to contain particles with diameters ranging from about 100 to 300 nanometers because this enhances their ability to strongly scatter light waves (Jovanović 2015). However, commercial TiO_2 ingredients are highly polydisperse systems that often contain a significant fraction of particles with diameters below 100 nm (Warheit et al. 2015). As an example, the mean particle diameter of food-grade TiO_2 (E171) powders from various ingredients manufacturers has been reported to be around 110 nm, with more than 36% of the particles having diameters less than 100 nm by number (Weir et al. 2012b). Consequently, people who consume large quantities of foods containing this additive may be exposed to high levels of inorganic nanoparticles. Indeed, it has been estimated that on average, humans are exposed to up to 1.1 and 2.2 mg/kg body weight of TiO_2 nanoparticles per day in the UK and US, respectively (Weir et al. 2012b). An important finding of this report was that the levels of TiO_2 nanoparticles consumed by children were around two- to four-fold greater than for adults. This effect can mainly be attributed to the fact that children tend to consume foods that contain relatively high concentrations of TiO_2 nanoparticles, like gums, sweets, desserts, and drinks.

In general, the TiO_2 nanoparticles within a food additive may have different crystalline forms, sizes, shapes, interfacial properties, and aggregation states (Figure 6.2), which influences their gastrointestinal fate and potential toxicity. Titanium dioxide may exist in different polymorphic forms in nanoparticles, with the anatase and rutile forms being the most common, which may influence their potential toxicity (Yang et al. 2014, Riediker et al. 2019). As mentioned earlier, the size of the particles in food-grade titanium dioxide ingredients may vary widely. In a single powdered ingredient, there may be particles with diameters ranging from tens to hundreds of nanometers (Geiss et al. 2021). Moreover, a large fraction of the individual particles may be agglomerated into relatively large clusters, which impacts their subsequent gastrointestinal fate. The surface chemistry of titanium dioxide particles can also vary depending on the method used to fabricate them as well as their interactions with other components in their environment (Yang et al. 2014). For instance, they may form biomolecular coronas in the gastrointestinal tract due to the adsorption of proteins and other substances on to their surfaces (Coreas et al. 2020, Sun et al. 2020, Pandit and Kundu 2021). All these factors may influence their gastrointestinal fate and potential to cause toxicity.

Studies where TiO_2 nanoparticles were orally administered to animals have shown that they may be absorbed and accumulate in their tissues, which may have adverse effects on health and wellbeing (Jovanović 2015). There have been hundreds of studies in this area, and only a few are highlighted here to provide some insights into the methodologies used. It has been reported that a single oral dose (5000 mg/kg body weight) of TiO_2 nanoparticles (25, 80, or 155 nm) led to their accumulation within the liver, kidneys, spleen, and lungs of mice as well as causing damage to their livers, kidneys, and hearts (Wang et al. 2007). Other researchers have reported that oral administration of higher doses (250 mg/kg body weight) of TiO_2 nanoparticles (5 nm) for 30 days caused damage to the liver and immune system of animals but lower doses (62.5 and 125 mg/kg body weight) did not (Duan et al. 2010). A recent study reported that oral administration of TiO_2 nanoparticles (0, 0.5, 5, or 50 mg/kg body weight per day) to female rats for 14 days led to accumulation of the nanoparticles within the animals tissues in a dose-dependent manner and had adverse effects on brain function (Canli et al. 2022). Oral administration of TiO_2 nanoparticles has also been reported to alter the composition of the gut microbiota, which could also have adverse effects on human health (Rinninella et al. 2021). In contrast, several researchers have reported little accumulation or toxicity of ingested TiO_2 nanoparticles. For instance, administration of different doses (260–1041 mg/kg body weight per day) of TiO_2 nanoparticles (21 nm) to rats was reported to cause no appreciable accumulation or toxicity (Cho et al. 2013). Instead, most of the nanoparticles were excreted in the feces of the animals.

There are several possible reasons for the apparently contradictory findings of different studies on the absorption and toxicity of TiO_2 nanoparticles in animal studies. First, there are differences in the concentration, size, shape, aggregation state, crystal form, and interfacial properties of the nanoparticles used in different studies (McClements et al. 2016b). Second, the influence of food matrix and gastrointestinal effects on nanoparticle characteristics and behavior are commonly not accounted for in these studies (McClements et al. 2016b, Mallia et al. 2022). Third, the type of animals and experimental methods utilized to establish nanoparticle absorption and toxicity vary. As an example, damage to the heart, liver, and stomach of young animals (3 weeks old) was observed after repeated oral administration of TiO_2 nanoparticles, but little toxicity was seen in adult animals (8 weeks old) (Wang et al. 2013c). This observation is particularly relevant because children tend to ingest higher amounts of TiO_2 nanoparticles than adults.

In vitro physicochemical and cell culture studies have been utilized to provide insights into the interactions of TiO_2 nanoparticles with other components under gastrointestinal conditions as well as to identify their potential absorption mechanisms and biochemical effects. *In vitro* experiments have shown that proteins and bile salts can bind to the surfaces of TiO_2 nanoparticles, thereby forming a biomolecular corona that alters their surface charge and aggregation state (Benbow et al. 2021, Korabkova et al. 2021, Yuan et al. 2022). Studies with cell culture models designed to simulate intestinal epithelium cells indicate that TiO_2 nanoparticles can be absorbed into the cells and promote cytotoxicity by an amount that depends on their concentration, size, shape, aggregation state, crystalline type, and interfacial properties (Gerloff et al. 2009, Chalew and Schwab 2013, Brun et al. 2014, Song et al. 2015, Tada-Oikawa et al. 2016). Other studies have shown that this type of nanoparticle may be absorbed without causing toxicity to the cells (Vila et al. 2018). Several mechanisms have been proposed to account for the toxic effects of ingested TiO_2 nanoparticles, such as the production of ROS that disrupt critical biochemical pathways in the cells, dysregulation of transporters and efflux pumps in the cell membranes, promotion of inflammation, damage to the mucus layer, and alterations in the microbiome in the colon (Kruger et al. 2014, Dorier et al. 2015a, Frohlich and Frohlich 2016, Limage et al. 2020). The crystalline state of the titanium dioxide within the nanoparticles has been reported to impact their toxicity. For instance, the anatase form was found to exhibit greater cytotoxicity than the rutile form, which was attributed to its stronger photo-catalytic activity (Dorier et al. 2015b). An important limitation of many of the cell culture studies in this area is that the cells are often exposed to an aqueous suspension of bare TiO_2 nanoparticles, without accounting for changes in their surface properties or aggregation state due to their interactions with components in foods or the gastrointestinal tract (Figure 6.2). It is well known that these kinds of interactions can alter nanoparticle absorption and toxicity in cell culture models (Monopoli et al. 2011, Lesniak et al. 2013).

The European Food Safety Authority (EFSA) Panel on Food Additives and Flavorings recently reviewed the safety data of titanium dioxide (E171) additives used in foods (Younes et al. 2021). The panel stated that <50% of the particles by number in E171 have diameters below 100 nm and so can be considered as nanoparticles. The panel reported that absorption of TiO_2 particles within the gastrointestinal tract is relatively low, but they may still accumulate inside the body. The impact of ingestion of either 1000 mg/kg body weight of E171 or 100 mg/kg body weight of TiO_2 nanoparticles (30 nm) per day on various toxicity markers was assessed. The panel reported that the ingestion of the TiO_2 particles did not lead to any significant reproductive or developmental toxicity effects, but there was some evidence for immunotoxicity, inflammation, genotoxicity, and neurotoxicity effects for some test subjects. However, there appeared to be no clear correlation between the physicochemical properties of

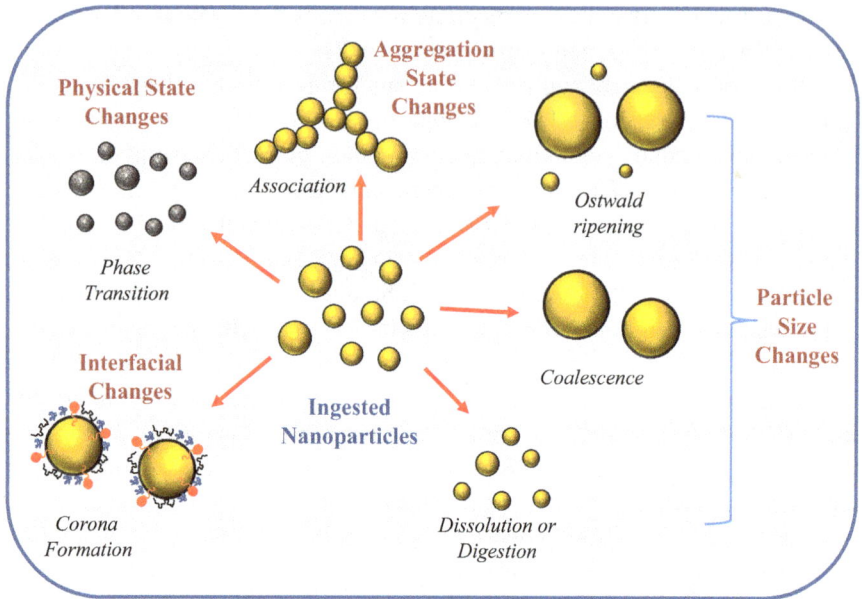

Figure 6.2: The properties of nanoparticles may change in several ways as they travel through the human gastrointestinal tract.

the TiO_2 particles and their impact on toxicity. Overall, the panel concluded that E171 could no longer be considered as safe when used as a food additive, based on the evidence of its potential toxicity but that more systematic studies were still required to obtain further evidence (Younes et al. 2021).

6.2.1.5 Silicon dioxide nanoparticles

Nanoparticles composed of SiO_2 are often used as additives in powdered foods (such as sugars, salts, spices, milks, and soups) because of their anticaking properties, i.e., the ability to inhibit clumping and improve flowability (Dekkers et al. 2011, Peters et al. 2012). These kinds of particles are typically solid amorphous spheres. Most of the particles in powdered SiO_2 food ingredients (E551) have diameters ranging from around 100 to 1000 nm, but there can also be an appreciable fraction of particles with smaller dimensions, which are therefore nanoparticles. The average ingestion of SiO_2 has been estimated to be about 20 to 50 mg/day per person (Frohlich and Frohlich 2016). Commercial SiO_2 powders have been reported to contain nanoparticles (10 to 50 nm) but these are often agglomerated into large clumps (100 to 1000 nm), which would be expected to influence their gastrointestinal fate (Yang et al. 2016). Several cell culture and animal feeding studies have reported that relatively high doses of SiO_2 nanoparticles may generate ROS and cause cytotoxicity (Athinarayanan et al. 2014, Yang et al. 2016). Oral administration of

SiO$_2$ nanoparticles to animals has been reported to lead to their accumulation within the liver, which could pose a health risk (van Kesteren et al. 2015). Indeed, some studies have reported that oral administration of SiO$_2$ nanoparticles to mice for 10 weeks adversely effected liver function (So et al. 2008). Conversely, other studies have reported little or no effects of ingesting this kind of nanoparticle. For instance, oral administration of silicon dioxide nanoparticles to rats for 13 weeks was reported to lead to no accumulation or toxicity (Yun et al. 2015). Again, these differences may be due to differences in the types of animals, silicon dioxide, and test methods used by different researchers.

The EFSA Panel on Food Additives and Nutrient Sources has evaluated the safety of silicon dioxide (E 551) as a food additive (Yun et al. 2015). They highlighted that several forms of silicon dioxide are used in foods and in toxicology studies, which are often not accurately characterized or reported. As a result, it is difficult to fully assess the risks associated with using nano-forms of silicon dioxide. Nevertheless, the existing evidence suggests that silicon dioxide appears to be poorly absorbed within the human gastrointestinal tract, although it can be found in some tissues (such as the liver and spleen) after oral ingestion. The evidence available at present suggests that silicon dioxide does not appear to have any adverse effects on human health at the levels utilized in foods, but there is still a need for further long-term studies on well-characterized silicon dioxide ingredients that might contain nanoparticles.

6.2.1.6 General comments

In general, there have been numerous studies on the absorption, tissue accumulation, and potential toxicity of inorganic nanoparticles. However, there are often apparently contradictory findings with different studies, with some of them reporting accumulation and/or toxicity and others finding no effects. There are several reasons for these apparent contradictions. The doses and kinds of nanoparticles utilized by different researchers in their studies vary considerably. Moreover, the levels of nanoparticles used in many cell culture and animal studies are typically much greater than the levels that are present within the human diet. *In vitro* digestion and cell culture models do not accurately simulate the complex nature of the gastrointestinal tract of humans, which means that the results of these studies should be treated with caution. The accumulation of inorganic nanoparticles within animal tissues is often determined using analytical methods that measure specific elements (like Ag, Zn, Fe, Ti, or Si) instead of measuring the presence of actual nanoparticles composed of these elements. Consequently, it is not clear whether it is the element itself or the nano-form that is leading to any observed toxicity. Finally, the impact of the food matrix that the nanoparticles are co-ingested with is often ignored. Studies have shown that the interaction of nanoparticles with food components (such as proteins, carbohydrates, lipids, or minerals) can greatly change their behavior by altering their surface characteristics or aggregation state (McClements and Xiao 2012). There is therefore a need for more *in vitro* and *in vivo* studies on the gastrointestinal behavior and toxicity of inorganic nanoparticles using well-characterized systems and standardized analytical methods.

6.2.2 Organic nanoparticles

Organic nanoparticles are fabricated from food-grade organic materials, like proteins, polysaccharides, phospholipids, saponins, and/or lipids (McClements 2021, Tan and McClements 2021). The core of these nanoparticles may be fluid, solid, or semi-solid depending on the ingredients and processing conditions used to formulate them as well as the prevailing environmental conditions (especially temperature). Moreover, solid organic nanoparticles may be either amorphous or crystalline, which impacts their encapsulation and release properties as well as their gastrointestinal fate. Organic nanoparticles may come in different sizes and shapes, with spheres and fibers being the most common in the food industry. As they move along the gastrointestinal tract, the properties of organic nanoparticles may be altered. For instance, they may be digested, dissolved, precipitated, aggregated, or disaggregated as they pass through the various regions of the gastrointestinal tract, such as the mouth, stomach, small intestine, or colon. Again, the gastrointestinal fate of this type of nanoparticle depends on their initial composition and structural organization. Usually, it is assumed that the toxicity of organic nanoparticles is less than that of inorganic ones. This is because many types of organic nanoparticles, such as those made from triacylglycerols, proteins, or starch, are fully digested in the upper gastrointestinal tract and therefore behavior similarly to other types of particles within the human gut. Other types of organic nanoparticles, such as those made from dietary fibers, are often fermented in the lower gastrointestinal tract, which may have health benefits. Even so, there are sometimes situations where organic nanoparticles could lead to potentially adverse health effects, which should be considered when formulating foods with them (see Section 6.5).

6.2.2.1 Lipid nanomaterials

Nanomaterials fabricated from edible lipids (such as triacylglycerol, flavor, or essential oils) are common in many food and beverage products (McClements 2013, McClements 2021). This type of nanoparticle is also being utilized to create delivery systems for the encapsulation, protection, and release of lipophilic active agents, e.g., oil-soluble vitamins, omega-3 fatty acids, flavors, colors, nutraceuticals, and preservatives (Livney 2015, Banasaz et al. 2020, Choi and McClements 2020). There are several potential benefits to utilizing lipid nanoparticles (rather than larger ones) in food and beverage applications: increased optical clarity (useful for applications in transparent foods and beverages); increased resistance to gravitational separation and aggregation (useful for applications where a long shelf life is required); increased digestibility and release (useful for applications where a high bioavailability is required) (Choi and McClements 2020). Lipid nanoparticles with different compositions and morphologies can be created from food-grade ingredients, including micelles, microemulsions, nanoliposomes, nanoemulsions, and solid lipid nanoparticles (Figure 6.3). These kinds of nanoparticles can have diameters that range from several nanometers (e.g., micelles) to several hundred nanometers (e.g., oil droplets). The interfacial characteristics of lipid nanoparticles (such as their charge, thickness, composition, and chemical reactivity) play an important role in determining their stability and behavior in foods and the gastrointestinal tract. It is therefore important to characterize the physicochemical and structural properties of lipid nanoparticles used in toxicology studies.

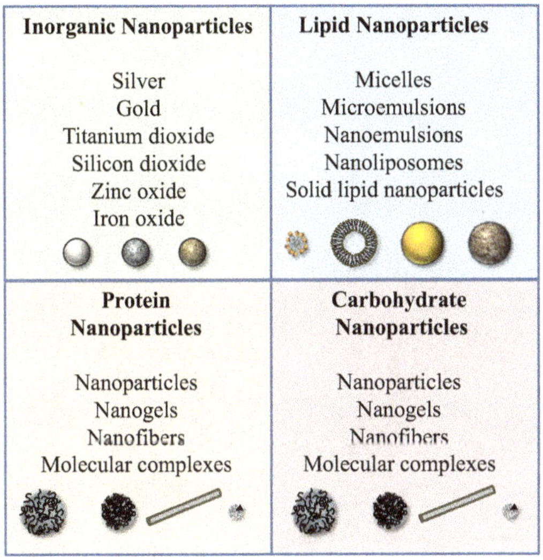

Inorganic Nanoparticles	Lipid Nanoparticles
Silver Gold Titanium dioxide Silicon dioxide Zinc oxide Iron oxide	Micelles Microemulsions Nanoemulsions Nanoliposomes Solid lipid nanoparticles
Protein Nanoparticles	**Carbohydrate Nanoparticles**
Nanoparticles Nanogels Nanofibers Molecular complexes	Nanoparticles Nanogels Nanofibers Molecular complexes

Figure 6.3: The gastrointestinal fate and potential toxicity of nanoparticles depend on their nature. Various kinds of inorganic or organic nanoparticles may be present within foods.

The behavior of lipid nanoparticles within the human gastrointestinal tract depends on digestibility by enzymes (lipases and phospholipases), which are mainly secreted in the small intestine and stomach (McClements and Xiao 2012, Amara et al. 2019). Lipid nanoparticles composed of triacylglycerols are normally converted into two free fatty acids and a monoacylglycerol by gastric and pancreatic enzymes. These lipid digestion products then combine with bile salts and phospholipids secreted within the small intestine to form colloidal structures known as mixed micelles, a natural form of lipid nanoparticle (McClements 2015b). The mixed micelles, which consist of a complex mixture of micelles and liposomes, are responsible for solubilizing and transporting lipophilic materials to the epithelium cells where they can be absorbed. The relatively high surface area of ingested lipid nanoparticles means they are rapidly and completely digested within the gastrointestinal tract. As a result, it would not be expected that this type of nanoparticle would be absorbed by the human body intact. Nevertheless, the rapid and complete digestion of this type of nanoparticle can increase the release and solubilization of encapsulated hydrophobic agents (like oil-soluble vitamins or nutraceuticals), thereby leading to an increase in their bioavailability. This is normally beneficial, but there may be unforeseen consequences for certain types of bioactive agents (e.g., nutrients that exhibit toxicity at high blood levels or undesirable hydrophobic substances like pesticides). The potential adverse effects of increasing the bioavailability of some ingested substances through this mechanism are discussed in Section 6.5 and highlighted in Figure 6.4.

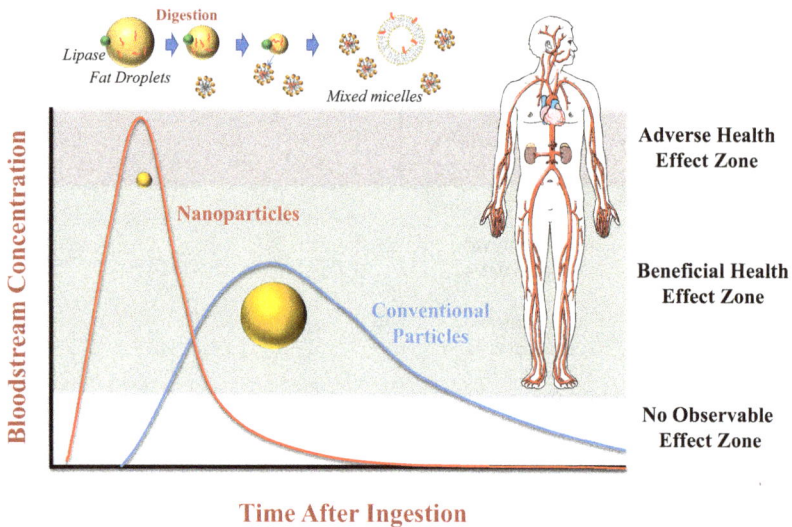

Figure 6.4: Schematic representation of change in the concentration of a bioactive substance in the bloodstream after ingestion. The size of the particles used to deliver a bioactive substance alter its release, transport, absorption, and metabolism within the gastrointestinal tract, thereby altering its concentration-time profile in the systemic circulation. The diagram of the human circulation system is from Servier Medical Art (Creative Commons 3.0).

Lipid nanoparticles made from indigestible oils (such as essential, flavor, or mineral oils) are not hydrolyzed by the digestive enzymes secreted by the human gastrointestinal tract (McClements and Rao 2011, McClements and Xiao 2012, Tan et al. 2019). As a result, they may behave differently from lipid nanoparticles made from digestible oils. For instance, they may absorb and retain oil-soluble nutrients and nutraceuticals, thereby reducing their absorption and bioactivity (Tan et al. 2019). If they remain sufficiently small within the gastrointestinal tract, they may be absorbed by the epithelium cells. Consequently, there is a need to better understand the gastrointestinal fate of indigestible lipid nanoparticles and their potential to cause adverse health effects.

6.2.2.2 Protein nanomaterials

The casein micelles found in milk are a natural form of edible nanoparticles since they have diameters of around 200 nm (Livney 2010, Verma et al. 2020). These nanoparticles are mainly composed of a mixture of different kinds of casein molecules (especially, α_{s1}, α_{s2}, β, and κ) held together by calcium phosphate complexes. Casein micelles have been consumed throughout human history as part of the milk obtained from humans, cows, sheep, goats, and other animals. Consequently, there is little concern about the potential health risks of consuming these types of edible nanoparticles. In fact, micellar casein probably evolved in mammals as a dispersible and digestible form of essential nutrients (such as amino acids and minerals) for growing infants (Oftedal 2012).

Many researchers are creating new kinds of protein nanoparticles that can be utilized in food products, e.g., as delivery systems, texture modifiers, or lightening agents (Davidov-Pardo et al. 2015, Livney 2015, Rajendran et al. 2016, Reddy and Rapisarda 2021). Inspired by nature, researchers are creating novel edible nanoparticles from other kinds of proteins. These nanoparticles typically have dimensions that range from several nanometers to several hundred nanometers depending on the type of proteins and fabrication methods used to prepare them. They can be assembled from animal proteins (such as casein, whey, ovalbumin, and gelatin), plant proteins (such as soy, pea, or potato proteins), or microbial proteins (such as algae proteins). Protein nanoparticles are typically small spheres, but other shapes are also possible, especially nanofibers. The surface charge of protein nanoparticles changes from negative to positive as the pH is reduced from above to below their isoelectric point (pI). The magnitude and sign of the charge on protein nanoparticles plays a major role in determining their stability and functionality. For instance, they tend to aggregate at pH values around the pI because of the reduction in electrostatic repulsion between them. Due to their relatively high surface areas, protein nanoparticles tend to be rapidly hydrolyzed within the upper gastrointestinal tract, thereby releasing peptides and amino acids. As a result, they would not be expected to have adverse health effects because they release similar digestion products as protein molecules or large aggregates after hydrolysis. It should be noted, however, that the types and amounts of bioactive peptides generated after digestion may be different for nanoparticles than for other forms. In principle, this could alter the allergenicity of the proteins but there does not appear to be any data in this area. Indigestible protein nanoparticles could be absorbed by the body, or they could interact with the gut microbiota, which could have some unforeseen consequences. Protein nanoparticles could also have some effects on human health by altering the bioavailability and/or bioactivities of encapsulated substances, such as minerals, vitamins, or nutraceuticals (Section 6.5).

6.2.2.3 Carbohydrate nanomaterials

Carbohydrate nanomaterials can be fabricated from various kinds of polysaccharides, including starch, cellulose, alginate, carrageenan, pectin, and their derivatives (Joye et al. 2014, Verma et al. 2020). They may come in various forms, including nanoparticles, nanogels, nanofibers, and molecular complexes (Figure 6.4). These kinds of nanoparticle may be produced using top-down or bottom-up methods. Top-down methods involve breaking down or dissociating relatively large natural substances, including starch granules, cellulose fibers, or chitosan fibers, into smaller substances, like nanocrystals or nanofibers. This can often be achieved using physical, chemical, and/or enzymatic methods. Bottom-up methods involve assembling multiple polysaccharide molecules into larger structures, such as nanoparticles or nanogels, which can often be achieved by adding crosslinking agents (such as mineral ions or enzymes) or by altering the temperature (such as heating or cooling). The nanoparticles created from polysaccharides are often spherical, but they may also be non-spherical, depending on the type of polysaccharides and fabrication methods used to produce them. Moreover, carbohydrate nanoparticles may be digestible (such as those formed from starch) or indigestible (such as those formed from cellulose, alginate, or carrageenan) within the upper gastrointestinal tract. Many starch-based nanoparticles can be digested due to the action of amylases secreted in the mouth and small intestine, which leads to the formation of oligosaccharides, disaccharides, and monosaccharides (Le Corre et al. 2010). However, other starch-based nanoparticles have structures that reduce their digestion within the upper gastrointestinal tract, such as those assembled from slowly digestible or resistant starch. Carbohydrate nanoparticles assembled from dietary fibers (such as cellulose, alginate, or carrageenan) tend to be resistant to digestion in the upper regions of the gastrointestinal tract (including the mouth, stomach, and small intestine). Even so, some of them may be dissembled as they pass through these regions due to weakening of the attractive forces holding them together. Moreover, nanoparticles fabricated from dietary fibers can be fermented by the enzymes secreted by the bacteria residing in the human colon. Consequently, the disintegration, digestion, and fermentation of this kind of nanoparticle have a major impact on their gastrointestinal fate and potential toxicity. If the carbohydrate nanoparticles are hydrolyzed within the upper gastrointestinal tract, they would not be expected to pose a major health risk because they generate similar digestion products (such as monosaccharides, disaccharides, and oligosaccharides) as regular forms of carbohydrates. Similarly, if the nanomaterials dissembled within the upper gastrointestinal tract, they would also be expected to behave like conventional carbohydrates. Conversely, if the carbohydrate nanoparticles are not dissembled or digested in the upper gastrointestinal tract, then they could be absorbed by the body or interact with the gut microbiota. As a result, they may behave in unanticipated ways that could lead to unforeseen health effects. For instance, small nanoparticles may be able to penetrate through the pores in the mucus layer coating the gastrointestinal tract and then be absorbed by the epithelium cells and accumulate in the body. Alternatively, some kinds of carbohydrate nanoparticles could have selective antimicrobial effects in the colon, which could alter the composition and function of the gut microbiome. Carbohydrate nanoparticles may also modulate the bioavailability of encapsulated bioactive agents, which could alter their health effects (Section 6.5).

6.2.2.4 Composite nanomaterials

Composite edible nanoparticles can be created by combining several kinds of different food ingredients, including proteins, polysaccharides, phospholipids, lipids, surfactants, and/or minerals (McClements 2015a). Nanoemulsions can be prepared that contain nanosized-oil droplets coated by single or multiple layers of biopolymers. Similarly, nanoliposomes can be prepared that contain phospholipid vesicles coated by one or more layer of biopolymers. Microgels can be formed that contain nanoemulsion droplets or solid lipid nanoparticles trapped inside. Nanogels and molecular complexes can be formed from oppositely charged proteins and polysaccharides due to electrostatic attraction. The utilization of different kinds of ingredients means that the compositional, structural, and functional properties of nanoparticles can be tailored for specific applications.

6.2.2.5 Gastrointestinal fate of organic nanoparticles

There have been relatively few systematic studies on the gastrointestinal behavior and toxicity of organic nanoparticles. One of the main hurdles to these kinds of studies is the lack of analytical instruments capable of characterizing the properties of these nanoparticles when they are dispersed within complicated food matrices or gastrointestinal fluids that consist of similar constituents (proteins, polysaccharides, phospholipids, and lipids). Consequently, it will be important in the future to develop and employ advanced analytical instrumentation to obtain a better understanding of the gastrointestinal fate of organic nanoparticles.

6.2.3 Plastic nanomaterials

There has been increasing concern about the presence of plastic nanoparticles and microparticles in foods and their potential toxicity (Ali et al. 2021, Stock et al. 2021, Kelpsiene et al. 2022). These particles are often the result of the degradation of plastics during their storage, such as polyethylene, polypropylene, polyethylene terephthalate, and polyvinyl chloride. For instance, nanoparticles and microplastics have been reported in seafood due to ingestion and accumulation of degraded plastics by marine animals (Piccardo et al. 2020). It has recently been reported that the size and composition of plastic particles impact their potential uptake by model epithelium cells but that they only exhibited cytotoxic effects when present at concentrations much higher than would normally be found in the human gastrointestinal tract (Stock et al. 2021). Clearly, more research is required in this area.

6.3 Characteristics of food nanoparticles

As discussed in earlier chapters, the organic and inorganic nanoparticles in foods and beverages have different properties (Figure 2.1), which would be expected to impact their gastrointestinal behavior and toxicity. It is therefore crucial when studying the potential toxicity of food nanoparticles to employ appropriate analytical instrumentation to characterize their composition, structure, dimensions, interfacial properties, and aggregation state (Wang et al. 2013b, Singh et al. 2014, Szakal et al. 2014, McClements and McClements 2016, More et al. 2021). Since the properties of food nanoparticles have been discussed in previous chapters, only a brief overview is given here, focusing on the link between these properties and their potential toxicity.

6.3.1 Composition

As discussed earlier, food nanoparticles can be composed of inorganic substances (such as silver, gold, ZnO, iron oxide, titanium dioxide, or silicon dioxide) or organic substances (such as carbon, lipids, phospholipids, proteins, and carbohydrates), as summarized in Figure 6.3. The composition of food nanoparticles influences their gastrointestinal fate and potential toxicity. Some kinds of ingested organic substances (such as lipids, proteins, and starches) used to assemble food nanoparticles are hydrolyzed by digestive enzymes (such as lipases, proteases, and amylases) within the upper gastrointestinal tract. But, other kinds (including mineral oils, resistant starch, and dietary fibers) are not digested, which impacts their potential absorption, accumulation, and distribution within the human body as well as their influence on the gut microbiome. Many kinds of ingested inorganic nanoparticles are not hydrolyzed by digestive enzymes and so may remain intact throughout the gastrointestinal tract. Nevertheless, some inorganic substances can dissolve in gastrointestinal fluids, which will alter their subsequent gastrointestinal fate (Dekkers et al. 2011, Frohlich and Frohlich 2016). Organic or inorganic nanoparticles not fully dissolved, digested, or absorbed within the mouth, stomach, and small intestine will reach the colon, where they can influence the composition and function of the gut microbiome (Williams et al.

2015, Frohlich and Frohlich 2016). The potentially toxic effects of inorganic nanoparticles are often linked to their chemical reactivity, such as their ability to generate ROS, which is at least partly governed by their compositions. For instance, silver nanoparticles can partially dissolve in the gastrointestinal fluids and release ions that can promote detrimental biochemical reactions, whereas titanium dioxide nanoparticles are relatively inert (Pradeep and Anshup 2009).

6.3.2 Dimensions

The external dimensions of the nanoparticles found in foods can range from several nanometers (e.g., micelles) to several hundred nanometers (e.g., titanium dioxide nanoparticles), which is governed by the materials and methods utilized in their assembly (Yada et al. 2014). The size of ingested nanoparticles impacts their fate within the gastrointestinal tract and therefore their potential to promote toxicity (Ensign et al. 2012, Frohlich and Roblegg 2012, Bellmann et al. 2015, Jain et al. 2018). Several physicochemical mechanisms may account for the impact of nanoparticle dimensions on their gastrointestinal fate and toxicity (Beloqui et al. 2016, Fan et al. 2016). First, dissolution or digestion of nanoparticles within gastrointestinal fluids usually increases as their size decreases because of the increase in their specific surface areas. Second, the interactions between ingested nanoparticles and gastrointestinal constituents (like enzymes, bile acids, and mineral ions) may increase as their size is reduced for the same reason. Third, the ability of nanoparticles to move through the layer of mucus that coats the epithelium cells in the gastrointestinal tract tends to increase as their dimensions decrease because they then can penetrate through the pores in the mucus layer more easily because these pores are typically a few hundred nanometers in size. Fourth, the transport of nanoparticles through (transcellular) or between (paracellular) epithelium cells depends on their dimensions. In particular, the cell membranes have active and passive transport mechanisms (like endocytosis) that depend on particle size. Consequently, a better understanding of the impact of nanoparticle dimensions on their gastrointestinal fate and potential toxicity is required. It should be noted that the dimensions of nanoparticles may change as they pass through the gastrointestinal tract due to processes such as dissolution, digestion, or aggregation (Figure 6.2), which affects their potentially adverse health effects.

6.3.3 Interfacial properties

The gastrointestinal behavior and potential toxicity of ingested nanoparticles also depend on the characteristics of their interfaces, such as their composition, thickness, and charge (Weir et al. 2012a, McClements et al. 2016a, McClements and Xiao 2017, Deng et al. 2021, Ghebretatios et al. 2021). Immediately after their production, nanoparticles may have a certain interfacial composition. However, after exposure to food matrices and gastrointestinal fluids, their interfacial properties may change due to adsorption of other substances in their environment, such as proteins, polysaccharides, phospholipids, surfactants, or mineral ions (Figure 6.2). The formation of this biomolecular corona often causes changes in the behavior of the nanoparticles, with the extent of these changes depending on its composition, thickness, polarity, charge, digestibility, and chemical reactivity. For instance, these interfacial properties may alter the aggregation state of nanoparticles in the gastrointestinal tract as well as their ability to move through mucus layers or epithelium cells. Further research is therefore needed to understand the formation and consequences of these biomolecular coronas.

6.3.4 Aggregation state

In food matrices and the human gastrointestinal tract, nanoparticles can be present as single particles or as clusters of particles (Figure 6.2), which may differ in their dimensions, structural organization, and resistance to disruption. Clustered nanoparticles are typically held together by attractive physical interactions, including van der Waals, electrostatic, and hydrophobic interactions. As a result, the aggregation state is sensitive to the conditions of the surrounding fluids, especially the temperature, pH, ionic composition, bridging interactions, and applied shear stresses. Nanoparticles that form clusters behave differently in the gastrointestinal tract than those that exist as individual entities, thereby influencing their gastrointestinal fate and potential toxicity. For instance, large clusters would not be expected to travel through gastrointestinal fluids, mucus layers, or epithelium cells as easily as the individual particles. For this reason, it is crucial to understand changes in the aggregation state of nanoparticles throughout the gastrointestinal tract.

6.4 Influence of food matrix and gastrointestinal effects on nanoparticle behavior

When considering the potential toxicity of ingested nanoparticles, it is important to account for changes in their properties due to interactions with constituents in their environments, such as food matrices and gastrointestinal fluids (Bellmann et al. 2015, McClements et al. 2016b). Prior to ingestion, the nanoparticles in foods and beverages are surrounded by different kinds of molecules, such as proteins, polysaccharides, phospholipids, surfactants, phytochemicals, and mineral ions, that can adsorb to their surfaces and alter their behaviors. Moreover, they may be present in fluids that have certain solution conditions, such as pH, ionic strength, and temperature. After ingestion, the nanoparticles are dispersed within gastrointestinal fluids that also contain a range of additional molecules that can also adsorb to their surfaces, including digestive enzymes, bile salts, phospholipids, and mucin. In addition, the pH, ionic composition, and temperature of these fluids may be different from those of the original food or beverage matrix. Changes in solution conditions and the adsorption of additional molecules to their surfaces alter the interfacial characteristics and aggregation state of nanoparticles, which alters their gastrointestinal fate, absorption, and potential toxicity (Figure 6.2).

Some of the important factors that may alter the behavior of ingested nanoparticles in the gastrointestinal tract are briefly discussed below (McClements and Xiao 2017):

pH and ionic strength: The pH and ionic strength of the gastrointestinal fluids are governed by the characteristics of the foods consumed but also by the gastrointestinal environment. Typically, the mouth, small intestine, and colon are fairly neutral or slightly acidic environments (pH 5.5 to 7.5), whereas the stomach is strongly acidic (pH 1 to 3). The pH and ionic composition of the gastrointestinal fluids influences the surface charge of ingested nanoparticles, which modifies their attractive and repulsive electrostatic interactions, thereby impacting their aggregation state and interactions with other components in the gastrointestinal tract.

Surface-active substances: Surface-active substances in the gastrointestinal fluids surrounding ingested nanoparticles may adsorb to their surfaces and alter their interfacial properties and interactions, thereby modifying their gastrointestinal behavior and toxicity. These surface-active substances may come from foods or from the gastrointestinal secretions, which include proteins, polysaccharides, phospholipids, surfactants, free fatty acids, and bile acids.

Enzyme activity: The human gastrointestinal tract contains various kinds of digestive enzymes, including lipases, proteases, and amylases, which can digest macronutrients such as lipids, proteins, and starches. These enzymes can therefore hydrolyze organic nanoparticles constructed from these substances. Alternatively, the nanoparticles may alter the activity of the enzymes, thereby interfering with normal digestion processes.

Biopolymers: Different kinds of biopolymers, such as polysaccharides, proteins, and glycoproteins, may be present within the fluids that surround nanoparticles in the human gastrointestinal tract. These biopolymers may have been present in the food or beverage that was consumed with the nanoparticles or they may have been secreted by the gastrointestinal tract. Biopolymers may have a variety of effects on the behavior of ingested nanoparticles. For instance, they may change the rheology of the gastrointestinal fluids (such as the viscosity), thereby altering the transport of the nanoparticles. They may promote or oppose the aggregation of nanoparticles within the gastrointestinal tract depending on the nature of the interactions involved. Finally, they may adsorb to the surfaces of the nanoparticles and alter their interfacial properties. All these factors could therefore influence the gastrointestinal fate and toxicity of the ingested nanoparticles.

Gastrointestinal barriers: The human gastrointestinal tract consists of epithelium cells that are coated by a structurally and compositionally complex mucus layer (Ensign et al. 2012, Carlson et al. 2018). Before they can be absorbed into the human body, the nanoparticles must penetrate through the mucus layer and epithelium cells. The interactions of the nanoparticles with these gastrointestinal barriers depend on their dimensions, interfacial properties, charge, and aggregation state within the appropriate region of the gastrointestinal tract.

Gut microbiome: The colon contains most of the microorganisms that reside in the human gastrointestinal tract. The type, number, and function of these microorganisms play a critical role in human health and wellbeing. Consequently, it is important to understand the potential impact of these microorganisms on nanoparticle behavior as well as the potential influence of ingested nanoparticles on the gut microbiome (Jiang et al. 2018, Ghebretatios et al. 2021). The bacteria in the gut microbiome secrete enzymes and other substances that can alter the properties and behavior of nanoparticles that reach the colon. For instance, bacteria may secrete enzymes that can ferment organic nanoparticles composed of dietary fibers that are indigestible in the upper gastrointestinal tract. This may be important for developing colon-specific delivery systems or for improving gut health. On the other hand, ingested nanoparticles reaching the colon could alter the type and number of bacteria in the gut microbiome, which could have potentially adverse health implications (Frohlich and Frohlich 2016, Jiang et al. 2018, Lamas et al. 2020). For instance, some indigestible nanoparticles exhibit antimicrobial activity, which could change the balance of microorganisms living within the colon, thereby having unknown health effects (Frohlich and Frohlich 2016, Pietroiusti et al. 2016, Li et al. 2019). It has also been reported that some kinds of inorganic nanoparticles can influence the bioactivity of other substances (like antibiotics), which has been attributed to their ability to disrupt the cell membranes of the gut microbes (Das et al. 2016).

Mechanical stresses: Nanoparticles are subjected to different types of stresses and fluid flow profiles as they pass through the gastrointestinal tract because of the mechanical action of the mouth (mastication), stomach (churning), small intestine (peristalsis), and colon (peristalsis). These mechanical stresses may change the aggregation state of ingested nanoparticles by disrupting any clusters held together by weak forces. They may also influence the movement, mixing, and location of the nanoparticles within the gastrointestinal tract.

Overall effects: As mentioned earlier, the properties and behavior of nanoparticles in the gastrointestinal tract are impacted by the various factors mentioned above, which will influence their potential to cause toxicity. In particular, their composition, size, aggregation state, physical state, charge, and interfacial properties may be altered by these factors (Figure 6.2). Consequently, it is extremely important to account for these factors when designing experiments to elucidate the toxicity of nanoparticles. As an example, studies have shown that ingesting inorganic nanoparticles with bread increases their absorption by epithelium cells, highlighting the importance of considering food matrix effects (Di Silvio et al. 2015). In general, it is therefore important to consider the properties of the nanoparticles within the gastrointestinal tract after they are ingested with specific foods rather than just focusing on the properties of the original nanoparticles in isolation before ingestion. If this is not done, then the results of toxicology studies may be unrealistic and misleading.

6.5 Potential causes of nanoparticle toxicity

Nanoparticles within the human gastrointestinal tract could lead to potentially adverse health effects through a variety of mechanisms. The relative importance of these mechanisms is likely to be influenced by the composition, dimensions, interfacial properties, and aggregation state of the nanoparticles. Several of the most important potential causes of nanoparticle toxicity are highlighted within this section.

6.5.1 Alterations in gastrointestinal function

High doses of certain kinds of ingested nanoparticles could cause alterations in normal gastrointestinal operations, such as digestion, transport, metabolism, and absorption (McClements and Xiao 2017). Nanoparticles have a relatively large specific surface area due to their small dimensions. Consequently, there is a large surface area for any surface-active substances in the gastrointestinal fluids to adsorb to, including enzymes, bile salts, mucin, and other substances. Most digestive and metabolic enzymes are globular proteins, whose catalytic activity depends on their precise three-dimensional conformation. Globular proteins often undergo conformational changes after adsorbing to the surfaces of particles, thereby leading to surface denaturation and a decrease in catalytic activity. In principle, high concentrations of nanoparticles in the gastrointestinal tract could therefore reduce the activity of amylases, lipases, and proteases, which could inhibit the digestion of starches, lipids, and proteins. However, the concentration of nanoparticles present is usually so low that this mechanism does not play a major role in macronutrient digestion. Moreover, other factors may impact the adsorption of enzymes to indigestible nanoparticles that reduce these effects: (i) the indigestible nanoparticles may aggregate, thereby decreasing their surface area; (ii) enzymes may adsorb more strongly to their substrates than to the surfaces of nanoparticles; (iii) other substances may also bind to the nanoparticle surfaces and reduce the space available for the enzymes to adsorb.

In the worst-case scenario, it would be anticipated that this mechanism would only lead to a modest decrease in the rate and extent of protein, starch, or lipid digestion. Moreover, the body may compensate by secreting extra digestive enzymes and other gastrointestinal constituents to ensure that most macronutrients were still fully digested. The concentrations of non-digestible nanoparticles typically found in the human diet are relatively low and so one would not anticipate this effect to cause significant health problems.

Ingested nanoparticles may also modulate macronutrient digestion by other mechanisms. For instance, certain kinds of inorganic nanoparticles may disrupt some of the barriers associated with the absorption of substances from the gastrointestinal tract, like the mucus layer, microvilli, or tight junctions, thus interfering with the uptake of macronutrients by the epithelium cells (Boegh and Nielsen 2015, Frohlich and Frohlich 2016). The existence of high concentrations of ingested nanoparticles within the gastrointestinal fluids could also trigger an immune response, which may also cause adverse health effects (Orfi and Szebeni 2016).

6.5.2 Tissue accumulation

There is evidence from studies on the oral administration of nanoparticles to animals that some of them may be taken up by the gastrointestinal tract and then accumulate in various tissues (Gaillet and Rouanet 2015, Ferdous and Nemmar 2020, de Souza et al. 2022). Presumably, ingested nanoparticles travel through the mucus layer and reach the surfaces of the epithelium cells where they are taken up by passive or active transport mechanisms. After being absorbed, they may be expelled back into the gastrointestinal tract, dissolve, be metabolized, accumulate inside the cells, or pass into the systemic circulation. The relative importance of these different processes is influenced by the properties of the nanoparticles, including their composition (digestible/non-digestible), size, shape, interfacial characteristics, and aggregation state. If nanoparticles accumulate within specific organs or tissues, they may cause cytotoxicity once they exceed a threshold concentration. This phenomenon is most likely to be an issue for inorganic nanoparticles that are not dissolved, metabolized, or digested within the cells where they accumulate.

6.5.3 Cytotoxicity

The presence of nanoparticles within organs and tissues may also promote cytotoxicity through other physicochemical mechanisms (Frohlich and Frohlich 2016, Sengul and Asmatulu 2020). Certain kinds of inorganic nanoparticles can produce ROS like singlet oxygen, superoxide, hydrogen peroxide, and hydroxyl radicals (Wu et al. 2014). These highly reactive substances can then promote cell damage by disrupting phospholipid membranes, proteins (such as enzymes, transporters, and signalers), and genetic materials (such as DNA and RNA) (Wu et al. 2014, Sengul and Asmatulu 2020, Yu et al. 2020). Consequently, some of the key biochemical pathways inside the cell needed to maintain their viability are adversely affected, including enzyme catalysis, energy generation, gene expression, and DNA replication (Sengul and Asmatulu 2020, Yu et al. 2020, Garcia-Torra et al. 2021). Moreover, certain kinds of nanoparticles can induce cytotoxicity due to their ability to dissolve and produce ions (like Ag^+ or Zn^{2+}) that can interact with cellular components (like lipids, proteins, or nucleic acids), thereby interfering with critical biochemical mechanisms. Inorganic nanoparticles are most likely to promote cytotoxicity because they can be absorbed by intestinal cells intact. Even so, there is still little information about the potential toxic effects of inorganic nanoparticles consumed at levels expected within the human diet and as part of compositionally and structurally complex food matrices.

6.5.4 Alteration in the site of release of bioactive substances

Encapsulating bioactive substances, such as vitamins, minerals, or nutraceuticals, in nanoparticles can change the site where they are released and absorbed in the human gastrointestinal tract. Nanoparticles can be designed to release bioactive substances within the mouth, stomach, small intestine, or colon, which may be different from the location they are normally released in. Consequently, the physiological response of the bioactive substances can be changed due to encapsulation in nanoparticles. In some cases, these changes could have undesirable health implications. For instance, encapsulating fats inside nanoparticles covered by dietary fiber coatings could reduce their digestion in the stomach and small intestine, thereby allowing them to reach the large intestine (McClements 2010). High levels of non-digested fat in the colon could cause gastrointestinal discomfort and other problems. As another example, encapsulating an antimicrobial substance (such as an essential oil) in nanoparticles that can reach the colon could cause changes in the number and type of microorganisms residing in the large intestine, which may have unforeseen health implications. Few studies have been carried out in this area to date, and clearly more research is required.

6.5.5 Enhancement of oral bioavailability

As discussed in previous chapters, a common use of nanotechnology within food science is to encapsulate, protect, and deliver hydrophobic bioactive substances, like oil-soluble vitamins, omega-3 oils, and nutraceuticals (Gorantla et al. 2021, Arshad et al. 2021, McClements 2021). Typically, the bioactive substances are trapped inside digestible nanoparticles, such as those composed of lipids, proteins, or starch, with lipids currently being the most widely used particle matrix for this purpose. Both *in vitro* and *in vivo* studies have demonstrated that the bioaccessibility and/or bioavailability of these bioactive substances tends to increase as the size of the particles they are trapped inside decreases. The main origin of this effect is that smaller particles have a larger specific surface area. As a result, they tend to be digested more rapidly and completely. This leads to a greater fraction of the bioactive substances being released into the surrounding gastrointestinal media. Moreover, in the case of lipid nanoparticles, this leads to the generation of more free fatty acids and monoacylglycerols that can be incorporated into mixed micelles, which can then solubilize and transport the released bioactive substances (McClements 2015c, McClements 2021). The increase in the digestion rate with decreasing particle size causes alterations in the pharmacokinetics of bioactive substances in the bloodstream (Figure 6.4). For instance, encapsulation of a hydrophobic bioactive substance within a digestible nanoparticle typically leads to a rapid spike in the bloodstream concentration, whereas encapsulation in larger particles leads to a more prolonged release profile. Alterations in the blood concentration levels of certain kinds of bioactive substances could have negative health implications. The biological activities of bioactive substances are typically governed by their concentrations in the bloodstream and tissues: they have little activity when the concentration is too low but may exhibit toxicity when the concentration is too high. Consequently, it is important that the concentration of the bioactive substances remains within some intermediate range for them to exhibit their potential beneficial activities. This optimum concentration range will depend on the substance being delivered as well as the physiology of the person ingesting the substance. Some bioactive substances exhibit little or no toxicity when ingested at relatively high concentrations and so would not be expected to cause problems if their bioavailability was greatly increased by encapsulation within nanoparticles. Conversely, the ability of nanoparticles to increase the bioavailability of other types of bioactive substances could have undesirable health consequences. Ingestion of some carotenoids (such as lutein and zeaxanthin) has been reported to improve human health, but ingestion of high levels of other carotenoids (such as beta-carotene) has been linked to increased cancer in some populations (such as smokers) (Black et al. 2020) Consequently, it is important to consider the potential impact of nanoencapsulation on the bioavailability of this kind of substance.

6.5.6 Increase in bioavailability of undesirable bioactive substances

Foods may contain both desirable (such as vitamins and nutraceuticals) and undesirable (such as pesticides) bioactive substances. *In vitro* gastrointestinal studies have shown that mixing hydrophobic pesticides with oil-in-water nanoemulsions increases the bioaccessibility of the pesticides (Zhang et al. 2017, Zhang et al. 2020). This effect would be expected to increase the amount of the pesticides absorbed by the human body, thereby having adverse health effects. The origin of this effect is that the lipid nanoparticles in the nanoemulsions are rapidly and completely digested by lipase in the gastrointestinal tract, thereby providing an increase in solubilization power of the mixed micelles due to the free fatty acids and monoacylglycerols generated.

6.5.7 Interference with gut microbiota

Any nanoparticles that can travel through the mouth, stomach, and small intestine will eventually reach the colon, where they can selectively impact the viability of the various species of bacteria residing there, thereby changing the composition and diversity of the gut microbiome (Frohlich and Frohlich 2016, Pietroiusti et al. 2016). The gut microbiome is known to have a major impact on human health and wellbeing (Xu and Knight 2015, Ghosh and Pramanik 2021). As a result, any alterations in the nature of the colonic bacteria caused by nanoparticles interacting with them may have unforeseen health implications (Ghebretatios et al. 2021). Further research is therefore required to better understand the influence of specific kinds of nanoparticles on the gut microbiome and the potential health implications.

6.6 Monitoring nanoparticle gastrointestinal fate and potential toxicity

Analytical instrumentation and experimental protocols are required to monitor the gastrointestinal fate of nanoparticles, establish the factors that impact their behavior under gastrointestinal conditions, and to determine their potential to cause toxicity. A variety of different approaches are typically required, ranging from *in vitro* laboratory assays to *in vivo* animal or human studies. The most important analytical instruments that are used to characterize nanoparticles were discussed in Chapter 2. In this section, a brief overview of some of the most common approaches used to establish the gastrointestinal fate and toxicity of nanoparticles is given. More detailed information can be found in review articles on this topic (Lefebvre et al. 2015, DeLoid et al. 2017, More et al. 2021). In addition, the EFSA agency has provided guidelines about the characteristics of nanoparticles that should be measured to assess their potential toxicity in foods, including their composition, size distribution, shape, structure, surface chemistry, melting point, boiling point, and density as well as different protocols that can be used to establish nanoparticle toxicity (More et al. 2021).

6.6.1 *In vitro* mechanistic studies

Information about the potential gastrointestinal fate of organic and inorganic nanoparticles, as well as the factors that impact them, can be obtained using *in vitro* digestion models that simulate the conditions in the mouth, stomach, small intestine, and colon (DeLoid et al. 2017). For instance, the residence times, flow profiles, pH values, ionic compositions, enzyme activities, mucus levels, bile salt concentrations, and microbial levels in the different regions of the gastrointestinal tract might be mimicked. The most widely used *in vitro* methods are designed to simulate the mouth (pH ≈ 7), stomach (pH ≈ 3), and small intestine (pH ≈ 7), but some methods also include the colon (pH ≈ 7). Typically, the nanoparticles are incubated sequentially in simulated oral, gastric, intestinal, and possibly colonic fluids and changes in their properties are measured, such as their composition, size, structure, shape, surface charge, location, and aggregation state. This knowledge can then be used to establish how specific gastrointestinal conditions impact nanoparticle behavior, such as the pH, flow profiles, mineral ion composition, digestive enzymes, bile salts, and mucin. Moreover, the nanoparticles can be mixed with different kinds of food matrices to examine the impact of specific food ingredients or structures on their behavior. Commonly, the nature of the biomolecular corona formed around the nanoparticles is measured in different food and gastrointestinal environments, as this impacts their interfacial properties and aggregation state, which in turn impacts their gastrointestinal behavior and potential toxicity.

6.6.2 Cell culture models

Cell culture models are widely used to simulate the epithelium cells that line the gastrointestinal tract, as this allows the factors that impact the absorption of nanoparticles into cells to be elucidated (DeLoid et al. 2017). In addition, they can be used to determine the fate of the nanoparticles inside the cells (such as their dissolution, accumulation, or expulsion) as well as their impact on key biochemical processes within the cells. Finally, they can be used to establish the potential of nanoparticles to cause cytotoxicity by measuring changes in cell viability.

Caco-2 cells, which are immortalized human colonic epithelial cells, are commonly used to simulate human gastrointestinal epithelial enterocytes because they have quite similar structures and functions. However, the epithelial enterocytes only make up part of the gastrointestinal barrier and so it is important for a more accurate presentation to include other kinds of cells in cell culture models. HT29-MTX cells can be included in these models because they resemble the goblet cells that secrete mucus in the gastrointestinal tract. The presence of a mucus layer is important because it acts as a semi-porous barrier that impacts the ability of nanoparticles to move from the intestinal fluids to the surfaces of the epithelium cells where absorption normally occurs. In addition, the small intestine also contains other kinds of cells, known as microfold or M-cells, that are capable of absorbing nanoparticles by engulfing and internalizing them. Consequently, it is also important to include these kinds of cells within cell culture models. Raji B cells have been used for this purpose because they can differentiate into cells that resemble the structure and function of M-cells. This triculture model is believed to provide a good representation of the human gastrointestinal tract and so can be used to examine the factors that impact the absorption of nanoparticles as well as their potential cytotoxicity (DeLoid et al. 2017).

6.6.3 *In vivo* animal models

In vitro digestion and cell culture models are extremely valuable for providing insights into the potential behavior of nanoparticles under simulated gastrointestinal conditions and can provide insights into the physicochemical mechanisms underlying this behavior. However, they cannot accurately simulate the extremely dynamic and complex physiological conditions occurring inside the human gastrointestinal tract. Studies with humans are difficult because there may be health risks, they are expensive, and they are time-consuming. For this reason, laboratory animals are often used to establish the gastrointestinal fate and toxicity of ingested nanoparticles. It is important to select laboratory animals that have gastrointestinal tracts that resemble those of humans, usually rodents or pigs. In these studies, the nanoparticles are usually orally administered to the animals, which can be as an aqueous dispersion in a drink, as part of a food matrix in the feed, or by oral gavage (More et al. 2021). In some situations, the behavior of the nanoparticles within the animal can be monitored non-invasively using imaging methods. However, more typically, their behavior is determined by collecting samples from different regions of the animal's gastrointestinal tract (such as stomach, small intestine, or colon) using inserted tubes or by sacrificing the animal. Nanoparticle characteristics such as their composition, size, shape, location, interfacial properties, charge, and aggregation state can then be measured using appropriate analytical methods (Chapter 2). By feeding animals different foods when they ingest the nanoparticles, the impact of food matrix effects on the gastrointestinal behavior of the nanoparticles can be elucidated. Moreover, the uptake of the nanoparticles in different organs and tissues of animals can be determined by scarifying them, harvesting the organs and tissues, and then analyzing them. Notably, this is typically done at only one or a limited number of time points after the animal has ingested the nanoparticles, so it is difficult to assess the location, properties, and effects of the nanoparticles within an animal's body over time. However, long-term feeding studies can be used to provide some insights into the organs where nanoparticles tend to accumulate.

There are some limitations with animal studies. The doses of nanoparticles used in these studies are often much higher than the nanoparticle concentrations found in the human diet, which limits the usefulness of the conclusions drawn. Moreover, the nanoparticles are often administered as an aqueous suspension or in a powdered form rather than being part of a complex food matrix, which also limits the utility of the findings because food components can alter the gastrointestinal fate and toxicity of nanoparticles.

6.6.4 Human studies

Human studies provide the most accurate information about the effects of ingested nanoparticles on health and wellbeing, but they also have many limitations. It is not ethical to feed humans nanoparticles expected to cause acute or chronic toxicity. The experiments are time-consuming and expensive. There are often large variations in the responses of different people, which leads to difficulties in obtaining results that are statistically significant. Moreover, only a limited number of parameters can be measured in human studies, such as the levels of the nanoparticles or their products in saliva, urine, feces, or blood. In some cases, imaging methods can be used to study the distribution of nanoparticles within the human body over time after ingestion (Munger et al. 2014). Moreover, biomarkers for potential adverse health effects can be measured in the blood and other tissues, such as the levels of ROS, pro-inflammatory cytokine generation, or enzyme activities (Munger et al. 2015).

6.7 Current status and future recommendations

There has been growing attention to establishing the potential toxicity of inorganic and organic nanoparticles in foods. These research efforts have provided important insights into the kinds of nanoparticles that may cause health concerns as well as the underlying physicochemical and physiological mechanisms that may impact their toxicity. Even so, there is still no consensus on the potential toxicity of different kinds of food-grade nanoparticles, which can be attributed to several causes:

- *Nanoparticle properties:* Toxicological studies often use nanoparticles that differ in their structural and physicochemical attributes, including their doses, compositions, sizes, shapes, aggregation states, physical states, and interfacial characteristics. Moreover, the attributes of the nanoparticles used in many studies are often not characterized or reported adequately, which makes it difficult to conclusively relate specific nanoparticle properties to their toxicity. A major concern is that the doses of nanoparticles used in many cell culture and animal studies are much higher than would ever be encountered as part of the human diet.
- *Characterization methods*: The potential toxicity of nanoparticles can be assessed using information obtained from different kinds of analytical tests, including *in vitro* physicochemical and cell culture models and *in vivo* animal and human studies, which vary in their level of sophistication. In addition, the analytical tests employed to characterize the nanoparticles, their behavior, and

their biological effects can differ appreciably from study-to-study, making it challenging to compare the results obtained from different studies.

– *Food matrix and gastrointestinal effects*: Many studies on the potential toxicity of nanoparticles ignore the impact of the food matrix and gastrointestinal tract on their behavior. For instance, pristine nanoparticles may be brought into direct contact with a cell culture model or may be orally administered directly to an animal. In practice, nanoparticles are typically embedded in complex food matrices or consumed with foods as part of a complex diet. Moreover, the nanoparticles must pass through the mouth, stomach, small intestine, and colon, which alter their properties and therefore potential toxicity. If food matrix and gastrointestinal effects are ignored, then the results obtained may not be realistic.

In future, it will be important to develop and employ standardized test procedures to establish the potential toxicity of nanoparticles that utilize more reproducible and realistic conditions.

6.8 Conclusions

As demonstrated throughout this book, the application of nanoparticles within the food industry is likely to increase due to their potential to enhance the quality, shelf-life, safety, and nutrition of foods as well as to provide novel effects (such as controlled or triggered release). Consequently, it is important to have a good understanding of their behavior within the gastrointestinal tract as well as their potential effects on human health and wellbeing. Moreover, food nanoparticles also have the potential to enter the environment and so it is also important to establish their potential impact on air, water, and soil quality as well as on the health of plants and animals. Nanoparticles are much smaller than the conventional particles used in our foods and so they may act differently inside our bodies and environment. Consequently, it is important to have knowledge of their behavior within the gastrointestinal tract, their accumulation within different tissues as well as their potential to cause adverse health effects. Currently, our understanding of the potential toxicity of most nanoparticles found in foods is still relatively limited and so it is difficult to make general recommendations about the safety of all types of food nanoparticles. Some nanoparticles are likely to be safe, whereas others are likely to be unsafe. It is therefore necessary to establish the potential toxicity of food nanoparticles on a case-by-case basis, which will be influenced by the type of nanoparticles and food matrix involved.

The mechanisms of action that are important in a particular situation depend on the nature of the nanoparticles used. For indigestible nanoparticles (like those made from titanium dioxide or cellulose), their potential absorption into the body, accumulation within specific tissues, and ability to promote cytotoxicity will be particularly important. In contrast, for digestible nanoparticles (like those made from some starches, lipids, or proteins), other factors may be more important. For instance, their rapid digestibility may lead to spikes in glucose, lipids, or amino acids in the bloodstream that could cause dysregulation of the metabolic system. Alternatively, the ability of these kinds of nanoparticles to greatly increase the bioaccessibility of substances in foods could increase the toxicity of undesirable food components (like hydrophobic pesticides or hormones), which could then lead to unforeseen health problems. Similarly, the ability of some digestible nanoparticles to greatly increase the bioavailability of nutrients that exhibit toxic effects when present at high concentrations in the body (like some fat-soluble vitamins and carotenoids) could also cause health concerns. It is clear that further research is still required to establish the potential magnitude and importance of these effects on human and environmental health.

References

Abbasi-Oshaghi, E., F. Mirzaei and A. Mirzaei (2018). "Effects of ZnO nanoparticles on intestinal function and structure in normal/high fat diet-fed rats and Caco-2 cells". Nanomedicine **13**(21): 2791–2816.

Ali, I., Q. H. Cheng, T. D. Ding, Y. G. Qian, Y. C. Zhang, H. B. Sun, C. S. Peng, I. Naz, J. Y. Li and J. F. Liu (2021). "Micro- and nanoplastics in the environment: Occurrence, detection, characterization and toxicity-A critical review". Journal of Cleaner Production **313**: 127863, 1–14. doi.org/10.1016/j.jclepro.2021.127863

Amara, S., C. Bourlieu, L. Humbert, D. Rainteau and F. Carriere (2019). "Variations in gastrointestinal lipases, pH and bile acid levels with food intake, age and diseases: Possible impact on oral lipid-based drug delivery systems". Advanced Drug Delivery Reviews **142**: 3–15.

Arshad, R., L. Gulshad, I. U. Haq, M. A. Farooq, A. Al-Farga, R. Siddique, M. F. Manzoor and E. Karrar (2021). "Nanotechnology: A novel tool to enhance the bioavailability of micronutrients". Food Science & Nutrition **9**(6): 3354–3361.

Athinarayanan, J., V. S. Periasamy, M. A. Alsaif, A. A. Al-Warthan and A. A. Alshatwi (2014). "Presence of nanosilica (E551) in commercial food products: TNF-mediated oxidative stress and altered cell cycle progression in human lung fibroblast cells". Cell biology and toxicology **30**(2): 89–100.

Bacchetta, R., E. Moschini, N. Santo, U. Fascio, L. Del Giacco, S. Freddi, M. Camatini and P. Mantecca (2014). "Evidence and uptake routes for Zinc oxide nanoparticles through the gastrointestinal barrier in Xenopus laevis". Nanotoxicology **8**(7): 728–744.

Banasaz, S., K. Morozova, G. Ferrentino and M. Scampicchio (2020). "Encapsulation of Lipid-Soluble Bioactives by Nanoemulsions". Molecules **25**(17): 3966, 1–21. doi.org/10.3390/molecules25173966

Bellmann, S., D. Carlander, A. Fasano, D. Momcilovic, J. A. Scimeca, W. J. Waldman, L. Gombau, L. Tsytsikova, R. Canady, D. I. A. Pereira and D. E. Lefebvre (2015). "Mammalian gastrointestinal tract parameters modulating the integrity, surface properties, and absorption of food-relevant nanomaterials". Wiley Interdisciplinary Reviews-Nanomedicine and Nanobiotechnology **7**(5): 609–622.

Beloqui, A., A. Des Rieux and V. Preat (2016). "Mechanisms of transport of polymeric and lipidic nanoparticles across the intestinal barrier". Advanced Drug Delivery Reviews **106**: 242–255.

Benbow, N. L., L. Rozenberga, A. J. McQuillan, M. Krasowska and D. A. Beattie (2021). "ATR FTIR Study of the Interaction of TiO2 Nanoparticle Films with beta-Lactoglobulin and Bile Salts". Langmuir **37**(45): 13278–13290.

Black, H. S., F. Boehm, R. Edge and T. G. Truscott (2020). "The Benefits and Risks of Certain Dietary Carotenoids that Exhibit both Anti- and Pro-Oxidative Mechanisms-A Comprehensive Review". Antioxidants **9**(3): 264, 1–19. https://doi.org/10.3390/antiox9030264.

Boegh, M. and H. M. Nielsen (2015). "Mucus as a Barrier to Drug Delivery – Understanding and Mimicking the Barrier Properties". Basic & Clinical Pharmacology & Toxicology **116**(3): 179–186.

Brun, E., F. Barreau, G. Veronesi, B. Fayard, S. Sorieul, C. Chaneac, C. Carapito, T. Rabilloud, A. Mabondzo, N. Herlin-Boime and M. Carriere (2014). "Titanium dioxide nanoparticle impact and translocation through ex vivo, in vivo and in vitro gut epithelia". Particle and Fibre Toxicology **11**: 13, 1–16. doi.org/10.1186/1743-8977-11-13

Bumbudsanpharoke, N. and S. Ko (2015). "Nano-Food Packaging: An Overview of Market, Migration Research, and Safety Regulations". Journal of Food Science **80**(5): R910–R923.

Canli, E. G., C. Gumus, M. Canli and H. B. Ila (2022). "The effects of titanium nanoparticles on enzymatic and non-enzymatic biomarkers in female Wistar rats". Drug and Chemical Toxicology **45**(1): 417–425.

Carlson, T. L., J. Y. Lock and R. L. Carrier (2018). "Engineering the Mucus Barrier". Annual Review of Biomedical Engineering Vol 20. M. L. Yarmush **20**: 197–220.

Cha, K., H. W. Hong, Y. G. Choi, M. J. Lee, J. H. Park, H. K. Chae, G. Ryu and H. Myung (2008). "Comparison of acute responses of mice livers to short-term exposure to nano-sized or micro-sized silver particles". Biotechnology letters **30**(11): 1893–1899.

Chalew, T. E. A. and K. J. Schwab (2013). "Toxicity of commercially available engineered nanoparticles to Caco-2 and SW480 human intestinal epithelial cells". Cell Biology and Toxicology **29**(2): 101–116.

Chen, N., Z. M. Song, H. Tang, W. S. Xi, A. N. Cao, Y. F. Liu and H. F. Wang (2016). "Toxicological Effects of Caco-2 Cells Following Short-Term and Long-Term Exposure to Ag Nanoparticles". International Journal of Molecular Sciences **17**(6): 974, 1–9. doi.org/10.3390/ijms17060974

Cho, W.-S., B.-C. Kang, J. K. Lee, J. Jeong, J.-H. Che and S. H. Seok (2013). "Comparative absorption, distribution, and excretion of titanium dioxide and zinc oxide nanoparticles after repeated oral administration". Particle and fibre toxicology **10**(1): 1.

Choi, S. J. and D. J. McClements (2020). "Nanoemulsions as delivery systems for lipophilic nutraceuticals: strategies for improving their formulation, stability, functionality and bioavailability". Food Science and Biotechnology **29**(2): 149–168.

Coreas, R., X. Q. Cao, G. M. DeLoid, P. Demokritou and W. W. Zhong (2020). "Lipid and protein corona of food-grade TiO2 nanoparticles in simulated gastrointestinal digestion". Nanoimpact **20**: 100272, 1–13. doi.org/10.1016/j.impact.2020.100272

Das, P., E. Saulnier, C. Carlucci, E. Allen-Vercoe, V. Shah and V. K. Walker. (2016). "Interaction between a Broad-spectrum Antibiotic and Silver Nanoparticles in a Human Gut Ecosystem". Journal of Nanomedicine & Nanotechnology **7**(5): 1–7.

Davidov-Pardo, G., I. J. Joye and D. J. McClements (2015). "Food-grade protein-based nanoparticles and microparticles for bioactive delivery: fabrication, characterization, and utilization". Advances in Protein Chemistry and Structural Biology **98**: 293–325.

de Souza, M. L., V. D. W. Sales, L. P. Alves, W. M. Dos Santos, L. R. D. Ferraz, G. S. D. Lima, L. M. D. Mendes, L. A. Rolim and P. J. R. Neto (2022). "A Systematic Review of Functionalized Polymeric Nanoparticles to Improve Intesti-nal Permeability of Drugs and Biological Products". Current Pharmaceutical Design **28**(5): 410–426.

Dekkers, S., P. Krystek, R. J. B. Peters, D. P. K. Lankveld, B. G. H. Bokkers, P. H. van Hoeven-arentzen, H. Bouwmeester and A. G. Oomen (2011). "Presence and risks of nanosilica in food products". Nanotoxicology **5**(3): 393–405.

DeLoid, G. M., Y. L. Wang, K. Kapronezai, L. R. Lorente, R. J. Zhang, G. Pyrgiotakis, N. V. Konduru, M. Ericsson, J. C. White, R. De La Torre-Roche, H. Xiao, D. J. McClements and P. Demokritou (2017). "An integrated methodology for assessing the impact of food matrix and gastrointestinal effects on the biokinetics and cellular toxicity of ingested engineered nanomaterials". Particle and Fibre Toxicology **14**: 40, 1–14. doi.org/10.1186/s12989-017-0221-5

Deng, J., Q. M. Ding, M. X. Jia, W. Li, Z. Zuberi, J. H. Wang, J. L. Ren, D. Fu, X. X. Zeng and J. F. Luo (2021). "Biosafety risk assessment of nanoparticles: Evidence from food case studies". Environmental Pollution **275**: 116662, 1–19. https://doi.org/10.1016/j.envpol.2021.116662.

Di Silvio, D., N. Rigby, B. Bajka, A. Mackie and F. B. Bombelli (2015). "Effect of protein corona magnetite nanoparticles derived from bread in vitro digestion on Caco-2 cells morphology and uptake". The international journal of biochemistry & cell biology 26520468, 1–9. doi.org/10.1016/j.biocel.2015.10.019

Dorier, M., E. Brun, G. Veronesi, F. Barreau, K. Pernet-Gallay, C. Desvergne, T. Rabilloud, C. Carapito, N. Herlin-Boime and M. Carriere (2015a). "Impact of anatase and rutile titanium dioxide nanoparticles on uptake carriers and efflux pumps in Caco-2 gut epithelial cells". Nanoscale **7**(16): 7352–7360.

Dorier, M., E. Brun, G. Veronesi, F. Barreau, K. Pernet-Gallay, C. Desvergne, T. Rabilloud, C. Carapito, N. Herlin-Boime and M. Carrière (2015b). "Impact of anatase and rutile titanium dioxide nanoparticles on uptake carriers and efflux pumps in Caco-2 gut epithelial cells". Nanoscale **7**(16): 7352–7360.

Duan, Y., J. Liu, L. Ma, N. Li, H. Liu, J. Wang, L. Zheng, C. Liu, X. Wang and X. Zhao (2010). "Toxicological characteristics of nanoparticulate anatase titanium dioxide in mice". Biomaterials **31**(5): 894–899.

Echegoyen, Y. and C. Nerin (2013a). "Nanoparticle release from nano-silver antimicrobial food containers". Food and Chemical Toxicology **62**: 16–22.

Echegoyen, Y. and C. Nerin (2013b). "Nanoparticle release from nano-silver antimicrobial food containers". Food and chemical toxicology: An international journal published for the British Industrial Biological Research Association **62**: 16–22.

EFSA (2016). "Safety assessment of the substance zinc oxide, nanoparticles, for use in food contact materials". EFSDA Journal **14**(3): 4408: 4401–4408.

Ensign, L. M., R. Cone and J. Hanes (2012). "Oral drug delivery with polymeric nanoparticles: The gastrointestinal mucus barriers". Advanced Drug Delivery Reviews **64**(6): 557–570.

Esmaeillou, M., M. Moharamnejad, R. Hsankhani, A. A. Tehrani and H. Maadi (2013). "Toxicity of ZnO nanoparticles in healthy adult mice". Environmental toxicology and pharmacology **35**(1): 67–71.

Fan, W. W., D. N. Xia, Q. L. Zhu, L. Hu and Y. Gan (2016). "Intracellular transport of nanocarriers across the intestinal epithelium". Drug Discovery Today **21**(5): 856–863.

Ferdous, Z. and A. Nemmar (2020). "Health Impact of Silver Nanoparticles: A Review of the Biodistribution and Toxicity Following Various Routes of Exposure". International Journal of Molecular Sciences **21**(7): 2375, 1–20. doi.org/10.3390/ijms21072375

Frohlich, E. and E. Roblegg (2012). "Models for oral uptake of nanoparticles in consumer products". Toxicology **291**(1–3): 10–17.

Frohlich, E. E. and E. Frohlich (2016). "Cytotoxicity of Nanoparticles Contained in Food on Intestinal Cells and the Gut Microbiota". International Journal of Molecular Sciences **17**(4): 509, 1–14. doi.org/10.3390/ijms17040509

Frtus, A., B. Smolkova, M. Uzhytchak, M. Lunova, M. Jirsa, S. Kubinova, A. Dejneka and O. Lunov (2020). "Analyzing the mechanisms of iron oxide nanoparticles interactions with cells: A road from failure to success in clinical applications". Journal of Controlled Release **328**: 59–77.

Fulgoni, V. L., D. R. Keast, R. L. Bailey and J. Dwyer (2011). "Foods, Fortificants, and Supplements: Where Do Americans Get Their Nutrients?". Journal of Nutrition **141**(10): 1847–1854.

Gaillet, S. and J. M. Rouanet (2015). "Silver nanoparticles: Their potential toxic effects after oral exposure and underlying mechanisms – A review". Food and Chemical Toxicology **77**: 58–63.

Garcia-Fernandez, J., D. Turiel, J. Bettmer, N. Jakubowski, U. Panne, L. R. Garcia, J. Llopis, C. S. Gonzalez and M. Montes-Bayon (2020). "In vitro and in situ experiments to evaluate the biodistribution and cellular toxicity of ultrasmall iron oxide nanoparticles potentially used as oral iron supplements". Nanotoxicology **14**(3): 388–403.

Garcia-Torra, V., A. Cano, M. Espina, M. Ettcheto, A. Camins, E. Barroso, M. Vazquez-Carrera, M. L. Garcia, E. Sanchez-Lopez and E. B. Souto (2021). "State of the Art on Toxicological Mechanisms of Metal and Metal Oxide Nanoparticles and Strategies to Reduce Toxicological Risks". Toxics **9**(8): 195, 1–13. doi.org/10.3390/toxics9080195.

Geiss, O., I. Bianchi, C. Senaldi, G. Bucher, E. Verleysen, N. Waegeneers, F. Brassinne, J. Mast, K. Loeschner, I. Vidmar, F. Aureli, F. Cubadda, A. Raggi, F. Iacoponi, R. Peters, A. Undas, A. Muller, A. K. Meinhardt, E. Walz, V. Graf and J. Barrero-Moreno (2021). "Particle size analysis of pristine food-grade titanium dioxide and E 171 in confectionery products: Interlaboratory testing of a single-particle inductively coupled plasma mass spectrometry screening method and confirmation with transmission electron microscopy". Food Control **120**: 107550, 1–18. doi.org/10.1016/j.foodcont.2020.107550

Georgantzopoulou, A., T. Serchi, S. Cambier, C. C. Leclercq, J. Renaut, J. Shao, M. Kruszewski, E. Lentzen, P. Grysan, S. Eswara, J. N. Audinot, S. Contal, J. Ziebel, C. Guignard, L. Hoffmann, A. J. Murk and A. C. Gutleb (2016). "Effects of silver nanoparticles and ions on a co-culture model for the gastrointestinal epithelium". Particle and Fibre Toxicology **13**: 9, 1–12. doi.org/10.1186/s12989-016-0117-9

Gerloff, K., C. Albrecht, A. W. Boots, I. Forster and R. P. F. Schins (2009). "Cytotoxicity and oxidative DNA damage by nanoparticles in human intestinal Caco-2 cells". Nanotoxicology **3**(4). 355–364.

Ghebretatios, M., S. Schaly and S. Prakash (2021). "Nanoparticles in the Food Industry and Their Impact on Human Gut Microbiome and Diseases". International Journal of Molecular Sciences **22**(4): 1942, 1–21. https://doi.org/10.3390/ijms22041942.

Ghosh, S. and S. Pramanik (2021). "Structural diversity, functional aspects and future therapeutic applications of human gut microbiome". Archives of Microbiology **203**(9): 5281–5308.

Gorantla, S., G. Wadhwa, S. Jain, S. Sankar, K. Nuwal, A. Mahmood, S. K. Dubey, R. Taliyan, P. Kesharwani and G. Singhvi "Recent advances in nanocarriers for nutrient delivery". Drug Delivery and Translational Research 1–15. doi.org/10.1007/s13346-021-01097-z

Hajipour, M. J., K. M. Fromm, A. A. Ashkarran, D. Jimenez de Aberasturi, I. Ruiz de Larramendi, T. Rojo, V. Serpooshan, W. J. Parak and M. Mahmoudi (2012). "Antibacterial properties of nanoparticles". Trends in Biotechnology **30**(10): 499–511.

Hendrickson, O. D., S. G. Klochkov, O. V. Novikova, I. M. Bravova, E. F. Shevtsova, I. V. Safenkova, A. V. Zherdev, S. O. Bachurin and B. B. Dzantiev (2016). "Toxicity of nanosilver in intragastric studies: Biodistribution and metabolic effects". Toxicology Letters **241**: 184–192.

Hilty, F. M., M. Arnold, M. Hilbe, A. Teleki, J. T. N. Knijnenburg, F. Ehrensperger, R. F. Hurrell, S. E. Pratsinis, W. Langhans and M. B. Zimmermann (2010). "Iron from nanocompounds containing iron and zinc is highly bioavailable in rats without tissue accumulation". Nature Nanotechnology **5**(5): 374–380.

Huang, Y. N., L. Ding, C. C. Li, M. Wu, M. M. Wang, C. J. Yao, X. L. Yin, J. F. Zhang, J. L. Liu, Y. Zhang, M. H. Wu and Y. L. Wang (2019). "Safety Issue of Changed Nanotoxicity of Zinc Oxide Nanoparticles in the Multicomponent System". Particle & Particle Systems Characterization **36**(10): 1900214, 1–9. doi.org/10.1002/ppsc.201900214

Jain, A., S. Ranjan, N. Dasgupta and C. Ramalingam (2018). "Nanomaterials in food and agriculture: An overview on their safety concerns and regulatory issues". Critical Reviews in Food Science and Nutrition **58**(2): 297–317.

Jeong, G. N., U. B. Jo, H. Y. Ryu, Y. S. Kim, K. S. Song and I. J. Yu (2010). "Histochemical study of intestinal mucins after administration of silver nanoparticles in Sprague-Dawley rats". Archives of toxicology **84**(1): 63–69.

Jiang, Z. Y., J. A. Jacob, J. Y. Li, X. H. Wu, G. L. Wei, A. Vimalanathan, R. Mani, P. Nainangu, U. M. Rajadurai and B. A. Chen (2018). "Influence of diet and dietary nanoparticles on gut dysbiosis". Microbial Pathogenesis **118**: 61–65.

Jovanović, B. (2015). "Critical review of public health regulations of titanium dioxide, a human food additive". Integrated environmental assessment and management **11**(1): 10–20.

Joye, I. J., G. Davidov-Pardo and D. J. McClements (2014). "Nanotechnology for increased micronutrient bioavailability". Trends in Food Science & Technology **40**(2): 168–182.

Jung, E. B., J. Yu and S. J. Choi (2021). "Interaction between ZnO Nanoparticles and Albumin and Its Effect on Cytotoxicity, Cellular Uptake, Intestinal Transport, Toxicokinetics, and Acute Oral Toxicity". Nanomaterials **11**(11): 2922, 1–15.

Kang, T. S., R. F. Guan, Y. J. Song, F. Lyu, X. Q. Ye and H. Jiang (2015). "Cytotoxicity of zinc oxide nanoparticles and silver nanoparticles in human epithelial colorectal adenocarcinoma cells". Lwt-Food Science and Technology **60**(2): 1143–1148.

Kelpsiene, E., M. T. Ekvall, M. Lundqvist, O. Torstensson, J. Hua and T. Cedervall (2022). "Review of ecotoxicological studies of widely used polystyrene nanoparticles". Environmental Science-Processes & Impacts **24**(1): 8–16.

Kim, T. H., M. Kim, H. S. Park, U. S. Shin, M. S. Gong and H. W. Kim (2012). "Size-dependent cellular toxicity of silver nanoparticles". Journal of biomedical materials research. Part A **100A**(4): 1033–1043.

Kim, Y. S., J. S. Kim, H. S. Cho, D. S. Rha, J. M. Kim, J. D. Park, B. S. Choi, R. Lim, H. K. Chang, Y. H. Chung, I. H. Kwon, J. Jeong, B. S. Han and I. J. Yu (2008). "Twenty-eight-day oral toxicity, genotoxicity, and gender-related tissue distribution of silver nanoparticles in Sprague-Dawley rats". Inhalation Toxicology **20**(6): 575–583.

Kim, Y. S., M. Y. Song, J. D. Park, K. S. Song, H. R. Ryu, Y. H. Chung, H. K. Chang, J. H. Lee, K. H. Oh, B. J. Kelman, I. K. Hwang and I. J. Yu (2010a). "Subchronic oral toxicity of silver nanoparticles". Particle and Fiber Toxicology **7**: 20.

Kim, Y. S., M. Y. Song, J. D. Park, K. S. Song, H. R. Ryu, Y. H. Chung, H. K. Chang, J. H. Lee, K. H. Oh, B. J. Kelman, I. K. Hwang and I. J. Yu (2010b). "Subchronic oral toxicity of silver nanoparticles". Particle and fibre toxicology **7**: 20.

Korabkova, E., V. Kasparkova, D. Jasenska, D. Moricova, E. Dadova, T. H. Truong, Z. Capakova, J. Vicha, J. Pelkova and P. Humpolicek (2021). "Behaviour of Titanium Dioxide Particles in Artificial Body Fluids and Human Blood Plasma". International Journal of Molecular Sciences **22**(19): 10614, 1–14. doi.org/10.3390/ijms221910614

Kruger, K., F. Cossais, H. Neve and M. Klempt (2014). "Titanium dioxide nanoparticles activate IL8-related inflammatory pathways in human colonic epithelial Caco-2 cells". Journal of Nanoparticle Research **16**(5): 2402, 1–13.

Lamas, B., N. M. Breyner and E. Houdeau (2020). "Impacts of foodborne inorganic nanoparticles on the gut microbiota-immune axis: potential consequences for host health". Particle and Fibre Toxicology **17**(1): 19, 1–19. doi.org/10.1186/s12989-020-00349-z

Le Corre, D., J. Bras and A. Dufresne (2010). "Starch Nanoparticles: A Review". Biomacromolecules **11**(5): 1139–1153.

Lefebvre, D. E., K. Venema, L. Gombau, L. G. Valerio, J. Raju, G. S. Bondy, H. Bouwmeester, R. P. Singh, A. J. Clippinger, E. M. Collnot, R. Mehta and V. Stone (2015). "Utility of models of the gastrointestinal tract for assessment of the digestion and absorption of engineered nanomaterials released from food matrices". Nanotoxicology **9**(4): 523–542.

Lesniak, A., A. Salvati, M. J. Santos-Martinez, M. W. Radomski, K. A. Dawson and C. Åberg (2013). "Nanoparticle adhesion to the cell membrane and its effect on nanoparticle uptake efficiency". Journal of the American Chemical Society **135**(4): 1438–1444.

Li, J. Y., M. Tang and Y. Y. Xue (2019). "Review of the effects of silver nanoparticle exposure on gut bacteria". Journal of Applied Toxicology **39**(1): 27–37.

Liang, T. S., R. F. Guan, M. Tao, F. Lyu, G. Z. Cao, M. Q. Liu and J. G. Gao (2017). "In Vitro Toxicity of Zinc Oxide Nanoparticles from Food Additives in Human Gastric Epithelium (GES-1) Cells". Science of Advanced Materials **9**(8): 1393–1400.

Lichtenstein, D., J. Ebmeyer, P. Knappe, S. Juling, L. Bohmert, S. Selve, B. Niemann, A. Braeuning, A. F. Thunemann and A. Lampen (2015). "Impact of food components during in vitro digestion of silver nanoparticles on cellular uptake and cytotoxicity in intestinal cells". Biological Chemistry **396**(11): 1255–1264.

Limage, R., E. Tako, N. Kolba, Z. Y. Guo, A. Garcia-Rodriguez, C. N. H. Marques and G. J. Mahler (2020). "TiO2 Nanoparticles and Commensal Bacteria Alter Mucus Layer Thickness and Composition in a Gastrointestinal Tract Model". Small **16**(21) e28939, 1–15. doi.org/10.5812/ircmj.28939

Livney, Y. D. (2010). "Milk proteins as vehicles for bioactives". Current Opinion in Colloid & Interface Science **15**(1–2): 73–83.

Livney, Y. D. (2015). "Nanostructured delivery systems in food: latest developments and potential future directions". Current Opinion in Food Science **3**: 125–135.

Loeschner, K., N. Hadrup, K. Qvortrup, A. Larsen, X. Y. Gao, U. Vogel, A. Mortensen, H. R. Lam and E. H. Larsen (2011). "Distribution of silver in rats following 28 days of repeated oral exposure to silver nanoparticles or silver acetate". Particle and Fibre Toxicology **8**: 18, 1–10. https://doi.org/10.1186/1743-8977-8-18.

Loza, K., C. Sengstock, S. Chernousova, M. Koller and M. Epple (2014). "The predominant species of ionic silver in biological media is colloidally dispersed nanoparticulate silver chloride". RSC Advances **4**(67): 35290–35297.

Mackevica, A., M. E. Olsson and S. F. Hansen (2016). "Silver nanoparticle release from commercially available plastic food containers into food simulants". Journal of Nanoparticle Research **18**(1): 5, 1–7. doi.org/10.1007/s11051-015-3313-x.

Mallia, J. D., R. Galea, R. Nag, E. Cummins, R. Gatt and V. Valdramidis (2022). "Nanoparticle Food Applications and Their Toxicity: Current Trends and Needs in Risk Assessment Strategies". Journal of Food Protection **85**(2): 355–372.

McClements, D. J. (2010). "Design of Nano-Laminated Coatings to Control Bioavailability of Lipophilic Food Components". Journal of Food Science **75**(1): R30–R42.

McClements, D. J. (2013). "Edible lipid nanoparticles: Digestion, absorption, and potential toxicity". Progress in Lipid Research **52**(4): 409–423.

McClements, D. J. (2015a). Nanoparticle- and Microparticle-based Delivery Systems. Boca Raton, FL, CRC Press.

McClements, D. J. (2015b). "Nanoscale Nutrient Delivery Systems for Food Applications: Improving Bioactive Dispersibility, Stability, and Bioavailability". Journal of Food Science **80**(7): N1602–N1611.

McClements, D. J. (2015c). "Reduced-fat foods: the complex science of developing diet-based strategies for tackling overweight and obesity". Advances in Nutrition **6**(3): 338S–352S.

McClements, D. J. (2021). "Advances in edible nanoemulsions: Digestion, bioavailability, and potential toxicity". Progress in Lipid Research **81**: 101081, 1–21. doi.org/10.1016/j. plipres.2020.101081

McClements, D. J., G. DeLoid, G. Pyrgiotakis, J. A. Shatkin, H. Xiao and P. Demokritou (2016a). "The role of the food matrix and gastrointestinal tract in the assessment of biological properties of ingested engineered nanomaterials (iENMs): State of the science and knowledge gaps". Nanoimpact **3–4**: 47–57.

McClements, D. J., G. Deloid, G. Pyrgiotakis, J. A. Shatkin, H. Xiao and P. Demokritou (2016b). "The role of the food matrix and gastrointestinal tract in the assessment ofbiological properties of ingested engineered nanomaterials (iENMs): Stateof the science and knowledge gaps". NanoImpact **3–4**: 47–57.

McClements, D. J. and J. Rao (2011). "Food-Grade Nanoemulsions: Formulation, Fabrication, Properties, Performance, Biological Fate, and Potential Toxicity". Critical Reviews in Food Science and Nutrition **51**(4): 285–330.

McClements, D. J. and H. Xiao (2012). "Potential biological fate of ingested nanoemulsions: influence of particle characteristics". Food & Function **3**(3): 202–220.

McClements, D. J. and H. Xiao (2017). "Is nano safe in foods? Establishing the factors impacting the gastrointestinal fate and toxicity of organic and inorganic food-grade nanoparticles". Npj Science of Food **1**(1): 6, 1–16. doi.org/10.1038/s41538-017-0005-1

McClements, J. and D. J. McClements (2016). "Standardization of Nanoparticle Characterization: Methods for Testing Properties, Stability, and Functionality of Edible Nanoparticles". Critical Reviews in Food Science and Nutrition **56**(8): 1334–1362.

Mittag, A., C. Hoera, A. Kampfe, M. Westermann, J. Kuckelkorn, T. Schneider and M. Glei (2021). "Cellular Uptake and Toxicological Effects of Differently Sized Zinc Oxide Nanoparticles in Intestinal Cells". Toxics **9**(5): 95, 1–14. doi.org/10.3390/toxics9050096

Monopoli, M. P., D. Walczyk, A. Campbell, G. Elia, I. Lynch, F. Baldelli Bombelli and K. A. Dawson (2011). "Physical– chemical aspects of protein corona: relevance to in vitro and in vivo biological impacts of nanoparticles". Journal of the American Chemical Society **133**(8): 2525–2534.

More, S., V. Bampidis, D. Benford, C. Bragard, T. Halldorsson, A. Hernandez-Jerez, S. H. Bennekou, K. Koutsoumanis, C. Lambre, K. Machera, H. Naegeli, S. Nielsen, J. Schlatter, D. Schrenk, V. Silano, D. Turck, M. Younes, J. Castenmiller, Q. Chaudhry, F. Cubadda, R. Franz, D. Gott, J. Mast, A. Mortensen, A. G. Oomen, S. Weigel, E. Barthelemy, A. Rincon, J. Tarazona, R. Schoonjans and E. S. Comm (2021). "Guidance on risk assessment of nanomaterials to be applied in the food and feed chain: human and animal health". Efsa Journal **19**(8): 6768, 1–70. https://doi.org/10.2903/j.efsa.2021.6768.

Munger, M. A., G. Hadlock, G. Stoddard, M. H. Slawson, D. G. Wilkins, N. Cox and D. Rollins (2015). "Assessing orally bioavailable commercial silver nanoparticle product on human cytochrome P450 enzyme activity". Nanotoxicology **9**(4): 474–481.

Munger, M. A., P. Radwanski, G. C. Hadlock, G. Stoddard, A. Shaaban, J. Falconer, D. W. Grainger and C. E. Deering-Rice (2014). "In vivo human time-exposure study of orally dosed commercial silver nanoparticles". Nanomedicine-Nanotechnology Biology and Medicine **10**(1): 1–9.

Oftedal, O. T. (2012). "The evolution of milk secretion and its ancient origins". Animal **6**(3): 355–368.

Orfi, E. and J. Szebeni (2016). "The immune system of the gut and potential adverse effects of oral nanocarriers on its function". Advanced Drug Delivery Reviews **106**: 402–409.

Pandit, S. and S. Kundu (2021). "Fluorescence quenching and related interactions among globular proteins (BSA and lysozyme) in presence of titanium dioxide nanoparticles". Colloids and Surfaces a-Physicochemical and Engineering Aspects **628**: 127253, 1–8. doi.org/10.1016/j.colsurfa.2021.127253

Parivar, K., F. M. Fard, M. Bayat, S. M. Alavian and M. Motavaf (2016). "Evaluation of Iron Oxide Nanoparticles Toxicity on Liver Cells of BALB/c Rats". Iranian Red Crescent Medical Journal **18**(1): e28939, 1–15. doi.org/10.5812/ircmj.28939

Park, E. J., E. Bae, J. Yi, Y. Kim, Y. Choi, S. H. Lee, J. Yoon, B. C. Lee and K. Park (2010). "Repeated-dose toxicity and inflammatory responses in mice by oral administration of silver nanoparticles". Environmental Toxicology and Pharmacology **30**(2): 162–168.

Pasupuleti, S., S. Alapati, S. Ganapathy, G. Anumolu, N. R. Pully and B. M. Prakhya (2012). "Toxicity of zinc oxide nanoparticles through oral route". Toxicology and industrial health **28**(8): 675–686.

Patil, U. S., S. Adireddy, A. Jaiswal, S. Mandava, B. R. Lee and D. B. Chrisey (2015). "In Vitro/In Vivo Toxicity Evaluation and Quantification of Iron Oxide Nanoparticles". International Journal of Molecular Sciences **16**(10): 24417–24450.

Peters, R., E. Kramer, A. G. Oomen, Z. E. H. Rivera, G. Oegema, P. C. Tromp, R. Fokkink, A. Rietveld, H. J. P. Marvin, S. Weigel, A. Peijnenburg and H. Bouwmeester (2012). "Presence of Nano-Sized Silica during In Vitro Digestion of Foods Containing Silica as a Food Additive". ACS Nano **6**(3): 2441–2451.

Piccardo, M., M. Renzi and A. Terlizzi (2020). "Nanoplastics in the oceans: Theory, experimental evidence and real world". Marine Pollution Bulletin **157**: 111317, 1–2.doi.org/10.1016/j.marpolbul.2020.111317

Pietroiusti, A., A. Magrini and L. Campagnolo (2016). "New frontiers in nanotoxicology: Gut microbiota/microbiome-mediated effects of engineered nanomaterials". Toxicology and Applied Pharmacology **299**: 90–95.

Pradeep, T. and Anshup (2009). "Noble metal nanoparticles for water purification: A critical review". Thin Solid Films **517**(24): 6441–6478.

Pulit-Prociak, J., K. Stoklosa and M. Banach (2015). "Nanosilver products and toxicity". Environmental Chemistry Letters **13**(1): 59–68: 31, 1–11. Edoi.org/10.3389/fchem.2016.00031.

Rajendran, S., C. C. Udenigwe and R. Y. Yada (2016). "Nanochemistry of Protein-Based Delivery Agents". Frontiers in Chemistry **4**

Raspopov, R. V., E. N. Trushina, I. V. Gmoshinsky and S. A. Khotimchenko (2011). "Bioavailability of nanoparticles of ferric oxide when used in nutrition. Experimental results in rats". Voprosy Pitaniya **80**(3): 25–30.

Reddy, N. and M. Rapisarda (2021). "Properties and Applications of Nanoparticles from Plant Proteins". Materials **14**(13).

Riediker, M., D. Zink, W. Kreyling, G. Oberdorster, A. Elder, U. Graham, I. Lynch, A. Duschl, G. Ichihara, S. Ichihara, T. Kobayashi, N. Hisanaga, M. Umezawa, T. J. Cheng, R. Handy, M. Gulumian, S. Tinkle and F. Cassee (2019). "Particle toxicology and health – where are we?". Particle and Fibre Toxicology **16**: 19, 1–14. doi.org/10.1186/s12989-019-0302-8

Rinninella, E., E. Cintoni, P. Raoul, V. Mora, A. Gasbarrini and M. C. Mele (2021). "Impact of Food Additive Titanium Dioxide on Gut Microbiota Composition, Microbiota-Associated Functions, and Gut Barrier: A Systematic Review of In Vivo Animal Studies". International Journal of Environmental Research and Public Health **18**(4): 2008, 1–19. doi.org/10.3390/ijerph18042008

Sengul, A. B. and E. Asmatulu (2020). "Toxicity of metal and metal oxide nanoparticles: a review". Environmental Chemistry Letters **18**(5): 1659–1683.

Shahare, B. and M. Yashpal (2013). "Toxic effects of repeated oral exposure of silver nanoparticles on small intestine mucosa of mice". Toxicology mechanisms and methods **23**(3): 161–167.

Sharifi, S., S. Behzadi, S. Laurent, M. L. Forrest, P. Stroeve and M. Mahmoudi (2012). "Toxicity of nanomaterials". Chemical Society Reviews **41**(6): 2323–2343.

Sharma, V. K., K. M. Siskova, R. Zboril and J. L. Gardea-Torresdey (2014). "Organic-coated silver nanoparticles in biological and environmental conditions: Fate, stability and toxicity". Advances in Colloid and Interface Science **204**: 15–34.

Singh, G., C. Stephan, P. Westerhoff, D. Carlander and T. V. Duncan (2014). "Measurement Methods to Detect, Characterize, and Quantify Engineered Nanomaterials in Foods". Comprehensive Reviews in Food Science and Food Safety **13**(4): 693–704.

Sirelkhatim, A., S. Mahmud, A. Seeni, N. H. M. Kaus, L. C. Ann, S. K. M. Bakhori, H. Hasan and D. Mohamad (2015). "Review on Zinc Oxide Nanoparticles: Antibacterial Activity and Toxicity Mechanism". Nano-Micro Letters **7**(3): 219–242.

So, S. J., I. S. Jang and C. S. Han (2008). "Effect of micro/nano silica particle feeding for mice". Journal of nanoscience and nanotechnology **8**(10): 5367–5371.

Song, Z. M., N. Chen, J. H. Liu, H. Tang, X. Y. Deng, W. S. Xi, K. Han, A. N. Cao, Y. F. Liu and H. F. Wang (2015). "Biological effect of food additive titanium dioxide nanoparticles on intestine: an in vitro study". Journal of Applied Toxicology **35**(10): 1169–1178.

Stock, V., C. Laurisch, J. Franke, M. H. Donmez, L. Voss, L. Bohmert, A. Braeuning and H. Sieg (2021). "Uptake and cellular effects of PE, PP, PET and PVC microplastic particles". Toxicology in Vitro **70**: 105021, 1–15. doi.org/10.1016/j.tiv.2020.105021

Sun, Y. J., T. Y. Zhen, Y. Li, Y. H. Wang, M. W. Wang, X. J. Li and Q. J. Sun (2020). "Interaction of food-grade titanium dioxide nanoparticles with pepsin in simulated gastric fluid". Lwt-Food Science and Technology **134**: 110208, 1–9. doi.org/10.1016/j.lwt.2020.110208

Szakal, C., S. M. Roberts, P. Westerhoff, A. Bartholomaeus, N. Buck, I. Illuminato, R. Canady and M. Rogers (2014). "Measurement of Nanomaterials in Foods: Integrative Consideration of Challenges and Future Prospects". ACS Nano **8**(4): 3128–3135.

Tada-Oikawa, S., G. Ichihara, H. Fukatsu, Y. Shimanuki, N. Tanaka, E. Watanabe, Y. Suzuki, M. Murakami, K. Izuoka, J. Chang, W. T. Wu, Y. Yamada and S. Ichihara (2016). "Titanium Dioxide Particle Type and Concentration Influence the Inflammatory Response in Caco-2 Cells". International Journal of Molecular Sciences **17**(4): 576, 1–11, doi.org/10.3390/ijms17040576

Tan, Y. and D. J. McClements (2021). "Plant-Based Colloidal Delivery Systems for Bioactives". Molecules **26**(22): 6895, 1–14. doi.org/10.3390/molecules26226895

Tan, Y. B., J. N. Liu, H. L. Zhou, J. M. Mundo and D. J. McClements (2019). "Impact of an indigestible oil phase (mineral oil) on the bioaccessibility of vitamin D-3 encapsulated in whey protein-stabilized nanoemulsions". Food Research International **120**: 264–274.

van Kesteren, P. C. E., F. Cubadda, H. Bouwmeester, J. C. H. van Eijkeren, S. Dekkers, W. H. de Jong and A. G. Oomen (2015). "Novel insights into the risk assessment of the nanomaterial synthetic amorphous silica, additive E551, in food". Nanotoxicology **9**(4): 442–452.

Vandebriel, R. J. and W. H. De Jong (2012). "A review of mammalian toxicity of ZnO nanoparticles". Nanotechnology, science and applications **5**: 61.

Verma, M. L., B. S. Dhanya, Sukriti, V. Rani, M. Thakur, J. Jeslin and R. Kushwaha (2020). "Carbohydrate and protein based biopolymeric nanoparticles: Current status and biotechnological applications". International Journal of Biological Macromolecules **154**: 390–412.

Vila, L., A. Garcia-Rodriguez, R. Marcos and A. Hernandez (2018). "Titanium dioxide nanoparticles translocate through differentiated Caco-2 cell monolayers, without disrupting the barrier functionality or inducing genotoxic damage". Journal of Applied Toxicology **38**(9): 1195–1205.

Voss, L., E. Hoche, V. Stock, L. Boehmert, A. Braeuning, A. F. Thuenemann and H. Sieg (2021). "Intestinal and hepatic effects of iron oxide nanoparticles". Archives of Toxicology **95**(3): 895–905.

Voss, L., I. L. Hsiao, M. Ebisch, J. Vidmar, N. Dreiack, L. Bohmert, V. Stock, A. Braeuning, K. Loeschner, P. Laux, A. F. Thunemann, A. Lampen and H. Sieg (2020). "The presence of iron oxide nanoparticles in the food pigment E172". Food Chemistry 327: 127000, 1–14. doi.org/10.1016/j.foodchem.2020.127000

Wang, B., W. Feng, M. Wang, T. Wang, Y. Gu, M. Zhu, H. Ouyang, J. Shi, F. Zhang and Y. Zhao (2008). "Acute toxicological impact of nano-and submicro-scaled zinc oxide powder on healthy adult mice". Journal of Nanoparticle Research 10(2): 263–276.

Wang, H., L.-J. Du, Z.-M. Song and -X.-X. Chen (2013a). "Progress in the characterization and safety evaluation of engineered inorganic nanomaterials in food". Nanomedicine 8(12): 2007–2025.

Wang, H., L. J. Du, Z. M. Song and X. X. Chen (2013b). "Progress in the characterization and safety evaluation of engineered inorganic nanomaterials in food". Nanomedicine (Lond) 8(12): 2007–2025.

Wang, J., G. Zhou, C. Chen, H. Yu, T. Wang, Y. Ma, G. Jia, Y. Gao, B. Li and J. Sun (2007). "Acute toxicity and biodistribution of different sized titanium dioxide particles in mice after oral administration". Toxicology letters 168(2): 176–185.

Wang, Y., Z. Chen, T. Ba, J. Pu, T. Chen, Y. Song, Y. Gu, Q. Qian, Y. Xu and K. Xiang (2013c). "Susceptibility of young and adult rats to the oral toxicity of titanium dioxide nanoparticles". Small 9(9-10): 1742–1752.

Wang, Y. L., L. L. Yuan, C. J. Yao, L. Ding, C. C. Li, J. Fang, K. K. Sui, Y. F. Liu and M. H. Wu (2014). "A combined toxicity study of zinc oxide nanoparticles and vitamin C in food additives". Nanoscale 6(24): 15333–15342.

Warheit, D. B., S. C. Brown and E. M. Donner (2015). "Acute and subchronic oral toxicity studies in rats with nanoscale and pigment grade titanium dioxide particles". Food and Chemical Toxicology 84: 208–224.

Weir, A., P. Westerhoff, L. Fabricius, K. Hristovski and N. von Goetz (2012a). "Titanium Dioxide Nanoparticles in Food and Personal Care Products". Environmental Science & Technology 46(4): 2242–2250.

Weir, A., P. Westerhoff, L. Fabricius, K. Hristovski and N. von Goetz (2012b). "Titanium dioxide nanoparticles in food and personal care products". Environmental science & technology 46(4): 2242–2250.

WHO (2000). Evaluation of national assessments of intake of iron oxides. Safety Evaluation of Certain Food Additives and Contaminants. Geneva, Switzerland, World Health Organization. 44: 1–2.

Williams, K., J. Milner, M. D. Boudreau, K. Gokulan, C. E. Cerniglia and S. Khare (2015). "Effects of subchronic exposure of silver nanoparticles on intestinal microbiota and gut-associated immune responses in the ileum of Sprague-Dawley rats". Nanotoxicology 9(3): 279–289.

Wu, H. H., J. J. Yin, W. G. Wamer, M. Y. Zeng and Y. M. Lo (2014). "Reactive oxygen species-related activities of nano-iron metal and nano-iron oxides". Journal of Food and Drug Analysis 22(1): 86–94.

Xu, Z. J. and R. Knight (2015). "Dietary effects on human gut microbiome diversity". British Journal of Nutrition 113: S1–S5.

Yada, R. Y., N. Buck, R. Canady, C. DeMerlis, T. Duncan, G. Janer, L. Juneja, M. Lin, D. J. McClements, G. Noonan, J. Oxley, C. Sabliov, L. Tsytsikova, S. Vazquez-Campos, J. Yourick, Q. Zhong and S. Thurmond (2014). "Engineered Nanoscale Food Ingredients: Evaluation of Current Knowledge on Material Characteristics Relevant to Uptake from the Gastrointestinal Tract". Comprehensive Reviews in Food Science and Food Safety 13(4): 730–744.

Yang, Y., K. Doudrick, X. Y. Bi, K. Hristovski, P. Herckes, P. Westerhoff and R. Kaegi (2014). "Characterization of Food-Grade Titanium Dioxide: The Presence of Nanosized Particles". Environmental Science & Technology 48(11): 6391–6400.

Yang, Y., J. J. Faust, J. Schoepf, K. Hristovski, D. G. Capco, P. Herckes and P. Westerhoff (2016). "Survey of food-grade silica dioxide nanomaterial occurrence, characterization, human gut impacts and fate across its lifecycle". Science of the Total Environment **565**: 902–912.

Younes, M., G. Aquilina, L. Castle, K. H. Engel, P. Fowler, M. J. F. Fernandez, P. Furst, U. Gundert-Remy, R. Gurtler, T. Husoy, M. Manco, W. Mennes, P. Moldeus, S. Passamonti, R. Shah, I. Waalkens-Berendsen, D. Waffle, E. Corsini, F. Cubadda, D. De Groot, R. FitzGerald, S. Gunnare, A. C. Gutleb, J. Mast, A. Mortensen, A. Oomen, A. Piersma, V. Plichta, B. Ulbrich, H. Van Loveren, D. Benford, M. Bignami, C. Bolognesi, R. Crebelli, M. Dusinska, F. Marcon, E. Nielsen, J. Schlatter, C. Vleminckx, S. Barmaz, M. Cart, C. Civitella, A. Giarola, A. M. Rincon, R. Serafimova, C. Smeraldi, J. Tarazona, A. Tard, M. Wright and E. P. F. A. Flavouri (2021). "Safety assessment of titanium dioxide (E171) as a food additive". Efsa Journal **19**(5): 6585, 1–70. doi.org/10.2903/j.efsa.2021.6585

Yu, Z. J., Q. Li, J. Wang, Y. L. Yu, Y. Wang, Q. H. Zhou and P. F. Li (2020). "Reactive Oxygen Species-Related Nanoparticle Toxicity in the Biomedical Field". Nanoscale Research Letters **15**(1): 115, 1–9. doi.org/10.1186/s11671-020-03344-7

Yuan, B. A., B. Jiang, H. Li, X. Xu, F. Li, D. J. McClements and C. J. Cao (2022). "Interactions between TiO2 nanoparticles and plant proteins: Role of hydrogen bonding". Food Hydrocolloids **124**: 107302, 1–12. doi.org/10.1016/j.foodhyd.2021.107302

Yun, J. W., S. H. Kim, J. R. You, W. H. Kim, J. J. Jang, S. K. Min, H. C. Kim, D. H. Chung, J. Jeong, B. C. Kang and J. H. Che (2015). "Comparative toxicity of silicon dioxide, silver and iron oxide nanoparticles after repeated oral administration to rats". Journal of Applied Toxicology **35**(6): 681–693.

Zhang, R. J., W. H. Wu, Z. P. Zhang, Y. Park, L. L. He, B. S. Xing and D. J. McClements (2017). "Effect of the Composition and Structure of Excipient Emulsion on the Bioaccessibility of Pesticide Residue in Agricultural Products". Journal of Agricultural and Food Chemistry **65**(41): 9128–9138.

Zhang, R. J., Z. P. Zhang, R. Y. Li, Y. B. Tan, S. S. Lv and D. J. McClements (2020). "Impact of Pesticide Type and Emulsion Fat Content on the Bioaccessibility of Pesticides in Natural Products". Molecules **25**(6): 1466, 1–9. doi.org/10.3390/molecules25061466.

Zimmermann, M. B. and F. M. Hilty (2011). "Nanocompounds of iron and zinc: their potential in nutrition". Nanoscale **3**(6): 2390–2398.

Index

www.ingramcontent.com/pod-product-compliance
Lightning Source LLC
Chambersburg PA
CBHW081526220326
41598CB00036B/6342